Classic Chemistry Demonstrations

Compiled by Ted Lister

Edited by Catherine O'Driscoll and Neville Reed

Designed by Imogen Bertin

Published by the Education Division, The Royal Society of Chemistry

Printed by The Royal Society of Chemistry

For further information on other educational activities undertaken by the Royal Society of Chemistry write to:

The Education Department
The Royal Society of Chemistry
Burlington House
Piccadilly
London W1V OBN

ISBN 1 870 343 38 7

British Library Cataloguing in Data.
A catalogue for this book is available from the British Library.

Classic Chemistry Demonstrations

Compiled by Ted Lister

Teacher Fellow
The Royal Society of Chemistry
1993–1994

This book is dedicated to the memory of

Jacqui Clee
Assistant Education Officer (Schools and Colleges)
The Royal Society of Chemistry 1991–1994

THE ROYAL
SOCIETY OF
CHEMISTRY

Foreword

Chemistry is an experimental subject, and what can be more stimulating than a chemical demonstration performed skilfully before an audience? I am sure that a majority of those who have been attracted to studying the subject have done so as a result of a memorable visual experiment performed in the laboratory or in the lecture theatre.

This collection of 100 chemistry demonstrations has been developed with the help and support of teachers throughout the United Kingdom and beyond. I hope that the experience represented in the collection will lead to a new generation being captured by the excitement of chemistry.

Professor J H Purnell OBE MA ScD EurChem CChem FRSC
President, The Royal Society of Chemistry

THE ROYAL
SOCIETY OF
CHEMISTRY

Contents

THE ROYAL
SOCIETY OF
CHEMISTRY

Introduction

Ted Lister

Since the curriculum development projects of the 1960s which stressed the importance of class practical work, demonstrations have gone somewhat out of fashion. However there are many reasons for carrying out demonstrations.

▼ They are often spectacular and therefore stimulating and motivating.

▼ They enable students to see experiments that they would not be able to perform themselves for a variety of reasons:

– the experiment requires chemical skills that are beyond their own;

– the experiment is potentially dangerous in unskilled hands;

– the experiment requires expensive apparatus and/or chemicals; and

– the experiment requires facilities, such as a fume cupboard, which are not available in sufficient numbers for class practical work.

▼ They allow students to see a skilled practitioner at work.

The ideas for the demonstrations in this book have come from a variety of sources. A few of them may be original, but most have been collected from the literature, both journals and books, (see the bibliography) and from chemistry teachers from all over the world. Many of the ideas come from more than one source and, in general, no attempt has been made to acknowledge the source. Variations on the "blue bottle" theme were suggested by at least 30 different people, for example!

Some of the demonstrations may be unfamiliar, others are classics. We have included many well known demonstrations because these will be useful to new chemistry teachers and to scientists from other disciplines who are teaching some chemistry.

All of the demonstrations have been tested at Warwick University by the author and have subsequently been trialled in schools and colleges where they have worked reliably. We had problems initially repeating some literature methods and often found that the temperature of the demonstration was critical. All of these experiments were carried out at room temperature (*ie* 15–25 °C) unless stated otherwise. Other factors can also affect the reproducibility of experiments, for example the exact degree of subdivision of powdered reagents. It is recommended that demonstrations are tried out thoroughly before doing them in front of an audience.

THE ROYAL
SOCIETY OF
CHEMISTRY

List of demonstrations

THE ROYAL
SOCIETY OF
CHEMISTRY

List of demonstrations by categories

Demonstration No.	1	2	3	4	5	6	7	8	9	10	11	12	13	14	15	16	17	18	19	20
Entertainment																				
Curriculum: Pre-16	✓	✓	✓		✓	✓	✓	✓			✓	✓		✓	✓		✓	✓	✓	
Post-16	✓	✓	✓	✓		✓	✓		✓	✓			✓	✓	✓	✓	✓	✓	✓	✓
Acid-base									✓	✓										
Electrochemistry				✓													✓			
Equilibria								✓	✓	✓										
Inorganic chemistry	✓		✓		✓						✓							✓		
Kinetics		✓										✓								
Organic chemistry													✓							✓
Periodicity															✓					
Polymers																				
Quantitative chemistry					✓		✓									✓			✓	
Thermodynamics						✓							✓							

Demonstration No.	21	22	23	24	25	26	27	28	29	30	31	32	33	34	35	36	37	38	39	40
Entertainment																				
Curriculum: Pre-16	✓	✓	✓		✓	✓	✓	✓	✓	✓	✓	✓	✓	✓	✓	✓	✓	✓	✓	
Post-16	✓			✓	✓	✓		✓	✓	✓	✓	✓		✓	✓	✓	✓	✓	✓	
Acid-base							✓												✓	
Electrochemistry																			✓	
Equilibria													✓							
Inorganic chemistry														✓						
Kinetics											✓									
Organic chemistry												✓								
Periodicity				✓																
Polymers	✓																			
Quantitative chemistry			✓																	
Thermodynamics								✓		✓					✓	✓	✓	✓		✓

THE ROYAL
SOCIETY OF
CHEMISTRY

Demonstration No.	41	42	43	44	45	46	47	48	49	50	51	52	53	54	55	56	57	58	59	60
Entertainment		✓					✓	✓				✓		✓						
Curriculum:																				
Pre-16	✓	✓	✓	✓	✓	✓	✓	✓	✓		✓	✓	✓	✓	✓	✓				✓
Post-16	✓	✓	✓	✓		✓	✓	✓		✓	✓	✓		✓			✓	✓	✓	✓
Acid-base												✓								
Electrochemistry			✓																	
Equilibria																				
Inorganic chemistry	✓					✓			✓		✓	✓	✓	✓	✓	✓			✓	
Kinetics					✓													✓		
Organic chemistry							✓	✓		✓										✓
Periodicity		✓							✓										✓	
Polymers																				
Quantitative chemistry													✓				✓			
Thermodynamics				✓										✓						

Demonstration No.	61	62	63	64	65	66	67	68	69	70	71	72	73	74	75	76	77	78	79	80
Entertainment		✓	✓	✓					✓										✓	
Curriculum:																				
Pre-16	✓	✓	✓	✓	✓	✓	✓	✓	✓	✓	✓	✓	✓	✓	✓	✓	✓	✓	✓	✓
Post-16		✓		✓	✓		✓	✓	✓	✓	✓						✓			✓
Acid-base																				
Electrochemistry										✓										
Equilibria																				
Inorganic chemistry	✓	✓	✓		✓	✓	✓	✓			✓	✓	✓	✓	✓	✓	✓	✓	✓	✓
Kinetics	✓	✓			✓				✓											
Organic chemistry				✓																
Periodicity												✓				✓	✓			
Polymers				✓																
Quantitative chemistry							✓							✓						
Thermodynamics								✓												✓

THE ROYAL
SOCIETY OF
CHEMISTRY

Demonstration No.	81	82	83	84	85	86	87	88	89	90	91	92	93	94	95	96	97	98	99	100
Entertainment			✓	✓						✓							✓		✓	
Curriculum:																				
Pre-16		✓	✓	✓	✓	✓	✓	✓	✓	✓	✓			✓	✓	✓	✓		✓	✓
Post-16	✓		✓	✓		✓	✓		✓	✓		✓	✓		✓	✓	✓			✓
Acid-base				✓	✓					✓										
Electrochemistry							✓							✓		✓				
Equilibria	✓																			
Inorganic chemistry			✓			✓			✓	✓		✓	✓				✓			✓
Kinetics		✓						✓												✓
Organic chemistry																		✓		
Periodicity						✓					✓									
Polymers															✓					
Quantitative chemistry																				
Thermodynamics																				

THE ROYAL
SOCIETY OF
CHEMISTRY

Safety

We believe that these chemical demonstrations can be carried out safely, but it is the responsibility of the teacher carrying out a particular demonstration to make the final decision depending on the circumstances at the time. Teachers must ensure that they follow the safety guidelines set down by their employers. A risk assessment must be completed for any demonstration that is carried out.

Below are the properties of some of the chemicals used in the demonstrations and are **highlighted** in the text.

Ammonia solution causes burns and gives off ammonia vapour which irritates eyes, lungs and respiratory system.

880 ammonia causes burns and gives off a vapour that irritates the eyes, lungs and respiratory system.

Ammonium dichromate decomposes violently when heated. It is an oxidising agent and assists fire. Dichromates irritate the skin, eyes and respiratory system.

Ammonium nitrate is a powerful oxidising agent, assists fire and can decompose explosively on strong heating.

Ammonium thiocyanate is harmful by inhalation, in contact with skin and if swallowed. It liberates a toxic gas on contact with concentrated acids or hot, dilute acids.

Barium chloride is a schedule 1 poison. It is harmful by inhalation and if swallowed.

Barium hydroxide is harmful by inhalation and swallowing. Wash the hands immediately with water in event of contact.

Bismuth oxide chloride is harmful if swallowed and the solid may irritate the eyes.

Bismuth trichloride is corrosive to the eyes and skin.

Bleach is an irritant.

Bromine vapour is an irritant and is very toxic if inhaled.

Calcium carbide liberates flammable ethyne gas on contact with water.

Chlorine is toxic. It is harmful to the eyes, lungs and respiratory tract.

Cobalt compounds are harmful if swallowed.

Copper compounds are harmful to the skin, eyes and lungs, and are toxic when ingested.

1, 2-dichloroethane is a carcinogen. It is highly flammable. It is harmful if swallowed and is irritating to the eyes, respiratory system and skin.

THE ROYAL
SOCIETY OF
CHEMISTRY

Cyclohexane is flammable and has a harmful vapour. It is irritating to skin and eyes.

Decanedioyl dichloride is corrosive.

1,6-diaminohexane is an irritant.

Glacial ethanoic acid gives off an irritating vapour. It is corrosive and causes severe burns.

Ethanol is flammable.

Ethylenediamine is flammable. It causes burns and is harmful in contact with the skin and if swallowed.

Formalin gives off an irritating vapour to which some people may be sensitive. It is toxic by inhalation, skin contact and if swallowed.

Hexane is flammable.

Hydrochloric acid can cause burns. It gives off an irritating vapour that can damage the eyes and lungs.

Concentrated hydrochloric acid causes burns. It gives off an irritating vapour which can damage the eyes and lungs.

Hydrogen is highly flammable. Ensure there are no naked flames in the laboratory. It forms explosive mixtures with air in concentrations of between 4 % and 74 %.

100 volume hydrogen peroxide is corrosive, irritating to the eyes and skin, and is a powerful oxidising agent.

Iodine is harmful by skin contact and gives off a toxic vapour that is dangerous to the eyes.

Iron(II) ethanedioate is harmful and an irritant.

Iron(III) nitrate is an oxidising agent that may assist fire.

Lead compounds are toxic.

Lead nitrate is an oxidising agent that assists fire.

Magnesium powder burns vigorously in air. The dust from magnesium powder may be hazardous.

Malonic acid irritates the eyes and skin.

Mercury has a toxic vapour. Work over a plastic spill tray and ensure that a suitable spillage kit is available.

THE ROYAL
SOCIETY OF
CHEMISTRY

Mercury compounds are very toxic by inhalation, skin contact and if swallowed. They cause cumulative effects.

Methane is flammable and forms explosive mixtures with air.

Nickel chloride-6-water is harmful and is an irritant.

Nitrates are powerful oxidising agents and assist fire.

Nitric acid causes severe burns and is an oxidising agent that may assist fire.

Concentrated nitric acid causes burns. Its vapour can damage eyes and lungs.

Nitrogen dioxide is toxic and irritates the respiratory system.

Oxygen supports combustion.

Liquid oxygen supports combustion.

Pentane is flammable and gives off a dense vapour.

Phenol is toxic by ingestion and skin absorption. It can cause severe burns. Take care when removing phenol from the bottle because the solid crystals can be hard to break up. Wear rubber gloves and a face mask.

Potassium bromate(V) is a strong oxidising agent and is toxic by ingestion.

Potassium chlorate is explosive when mixed with combustible material and is harmful by inhalation and if swallowed.

Potassium dichromate is a powerful oxidising agent. It can cause ulcers on contact with the skin and is a suspected carcinogen.

Potassium hydroxide is corrosive and can damage the eyes.

Potassium iodate is an oxidising agent and can assist fire.

Potassium permanganate is a skin irritant and is harmful if swallowed. It can cause fire on contact with combustible materials.

Propanone is flammable.

Silver nitrate causes burns. It is an oxidising agent and may assist fire.

Sodium is flammable and reacts with water to produce hydrogen and alkaline solutions.

Sodium chlorate is explosive when mixed with combustible material and is harmful by inhalation and if swallowed.

Sodium hydroxide can cause severe burns to the skin and is dangerous to the eyes.

Sodium hypochlorite solution (domestic bleach) is corrosive to the skin and eyes.

THE ROYAL
SOCIETY OF
CHEMISTRY

Sodium metabisulphite is irritating to the eyes, skin and respiratory system.

Sodium nitrate(III) is poisonous. It reacts with acids to give off oxides of nitrogen and will support combustion.

Sulphur is flammable and produces sulphur dioxide on burning in air.

Sulphur dioxide is an acidic and choking gas. It is toxic by inhalation and can affect asthma sufferers adversely.

Sulphuric acid is corrosive and causes burns.

Concentrated sulphuric acid is a strong acid, a powerful oxidant and a dehydrating agent. Its reaction with water is highly exothermic.

Toluene is highly flammable and harmful by inhalation.

1,1,1-trichloroethane is harmful by inhalation and its vapour irritates the eyes, skin and lungs.

Vanadium (V) salts are toxic by ingestion or by inhalation of dust. They irritate the skin.

Xylene is highly flammable and harmful by inhalation.

THE ROYAL
SOCIETY OF
CHEMISTRY

Acknowledgements

Ted Lister

I would like to thank a large number of people for their help in producing this book.

Neville Reed and the team in the Education Department at the Royal Society of Chemistry, Burlington House, for help, support and guidance.

Professor Keith Jennings in the Chemistry Department at Warwick University for providing accommodation and the staff, students and technicians in the Department for help in all sorts of ways.

Jill Bingham, Glynis Goodfellow, Pauline Lanham and Anne Wagg – the laboratory technicians at Trinity School, Leamington Spa for their patience and 'loan' of chemicals.

My wife (also a chemistry teacher), Jan.

The following people contributed ideas and/or trialled experiments:

Sally Baalham	Ipswich School
Elizabeth Barker	St Mary's School, Colchester
Martyn Berry	Chislehurst and Sidcup Grammar School, Kent
Sunil Bhalla	Trinity School, Leamington Spa
Grant Burleigh	Nailsea Comprehensive School, Nailsea
Oleg Chizhof	N. D. Zelinsky Institute of Organic Chemistry, Moscow
P. Clague	St Ninian's High School, Isle of Man
Richard Clarke	Northampton High School, Northampton
Mike Coles	Bristol Exploratory, Bristol
Philip Collie	The Cotswold School, Bourton-on-the-Water
Simon Cotton	Felixstowe College, Felixstowe
Julia Cox	St Edward's School, Oxford
David Curzons	St Christopher School, Letchworth
Jim Donnelly	University of Leeds, Leeds
Hugh Dunlop	American Community College, Cobham
Ann Fairmington	Dillyn Llewlyn Comprehensive Community School, Swansea
Ben Faust	Loughborough Grammar School, Loughborough
Chris Foyston	Austin Friars School, Carlisle
Tim Gayler	The Sanders Draper School, Hornchurch
Roger Gedge	Wellington College, Berkshire
Graham Hall	Halesowen College, Halesowen
Anne Hare	Merton Sixth Form College, Morden
Bridget Holligan	Bristol Exploratory, Bristol
Norman Hoyle	St Helen's School, Northwood
Mervyn Hudson	Chesterfield High School, Liverpool
Colin Johnson	Techniquest, Cardiff
Jan King	Broughton Hall High School, Liverpool
Ken Lang	Orpington, Kent
Gavin Lazaro	The Radcliffe School, Milton Keynes
Eric Lewis	Cranbook School, Kent
S. J. Lund	Long Crendon, Buckinghamshire
Judy Machin	Gumley House Convent School, Middlesex

Jenny Martin	Lode Heath School, Solihull
David Moore	St Edward's School, Oxford
Michael Morelle	Highgate School, London
Chris Noble	Bournemouth School for Girls, Bournemouth
Nick Parmar	Channing School, London
Mark Pilkington	Falmouth Community School, Falmouth
Ray Plevey	University of Birmingham, Birmingham
Jean Pocock	Newbury College of FE, Newbury
Bill Pritchard	University of Warwick, Warwick
Trevor Read	Finchley Catholic High School, Finchley
Brian Sanderson	The King John School, Benfleet
John Skinner	Tiffin Girls' School, Kingston
Andrew Szydlo	Highgate School, London
Dorothy Titchener	Tattenhall, Chester
Valerie Tordoff	Eton College, Berkshire
Colin Turner	Churchfields High School, West Midlands
K. Utting	St Bede's School, Redhill
Marjorie Walklett	Windsor, Berkshire
Elaine Wilson	Parkside Community College, Cambridge
George Wood	Sheffield College, Sheffield
Bob Worley	CLEAPSS, Brunel University

David Moore and his laboratory technician, Julia Cox at St Edward's School, Oxford, must be singled out for special mention, where over one third of the collection was trialled.

Michael Edenborough loaned a PGCE dissertation on demonstrations.

The Royal Society of Chemistry would like to thank the Chemistry Department at Warwick University for providing laboratory and office accommodation for this Fellowship, and the Head Teacher and Governors of Trinity School, Leamington Spa, and Warwickshire Local Education Authority for seconding Ted Lister to its Education Department.

THE ROYAL
SOCIETY OF
CHEMISTRY

Bibliography

H. N. Alyea and F. B Dutton (eds). *Tested demonstrations in chemistry.* Easton, Pennsylvania. Journal of Chemical Education, 1965.

G. Fowles, *Lecture experiments in chemistry.* London: Bell and Sons, 1959.

D. A. Humphreys, *Demonstrating chemistry.* Hamilton, Ontario: McMaster University, 1983.

B. Iddon, *The magic of chemistry.* Letchworth: Garden City Press, 1985.

B. Z. Shakhashiri, *Chemical demonstrations: a handbook for teachers of chemistry,* Vols 1–4. Madison: University of Wisconsin, 1983-92. More volumes are planned. These books are the 'Bible' of chemical demonstrations.

L. R. Summerlin, C. L. Borgford and J. L. Ealy, *Chemical demonstrations: a sourcebook for teachers,* Vol 2. Washington DC: American Chemical Society, 1988.

L. R. Summerlin and J. L. Ealy, *Chemical demonstrations: a sourcebook for teachers.* Washington DC: American Chemical Society, 1988.

I. Talesnick, *Idea bank collection.* Kingston, Ontario: Science Supplies and Services, 1991.

Tested Demonstrations in Chemistry can be found in each issue of *The Journal of Chemical Education* and ideas for demonstrations can often be found in *School Science Review* and *Education in Chemistry.*

More comprehensive annotated bibliographies, which include articles in journals, can be found in R. A. Schibeci, *Ed. Chem.,* 1988, p150 and B. Iddon, *School Sci. Rev.,* 1986, p704.

THE ROYAL
SOCIETY OF
CHEMISTRY

THE ROYAL
SOCIETY OF
CHEMISTRY

1. A visible activated complex

Topic

Reaction rates/catalysis.

Timing

About 5 min.

Level

Pre-16 as a simple demonstration of catalysis, post-16 to introduce the idea of an activated complex and to allow discussion of the mechanism of catalysis.

Description

Hydrogen peroxide oxidises potassium sodium tartrate (Rochelle salt) to carbon dioxide. The reaction is catalysed by cobalt(II) chloride. When solutions of hydrogen peroxide and Rochelle salt are mixed, carbon dioxide is slowly evolved. The addition of cobalt(II) chloride causes the reaction to froth, indicating a large increase in the reaction rate. At the same time the colour of the cobalt(II) chloride turns from pink to green (an activated complex), returning to pink again within a few seconds as the reaction dies down. This indicates that catalysts actually take part in the reaction and are returned unchanged when the reaction is complete.

Apparatus

▼ Bunsen burner, tripod, gauze and heat-proof mat.

▼ One 250 cm³ beaker.

▼ One 0 –100 °C thermometer.

▼ One 25 cm³ measuring cylinder.

▼ One dropping pipette.

▼ Access to overhead projector (optional).

Chemicals

The quantities given are for one demonstration.

▼ 5 g of potassium sodium tartrate-4-water (Rochelle salt, potassium sodium 2,3-dihydroxybutanedioate, $KNaC_4H_4O_6.4H_2O$).

▼ 0.2 g of **cobalt(II) chloride-6-water** ($CoCl_2.6H_2O$).

▼ 20 cm³ of 20 volume (*ie* approximately 6 %) hydrogen peroxide solution ($H_2O_2(aq)$).

▼ 65 cm³ of deionised water.

▼ About 200 cm³ of crushed ice (optional).

Method

Before the demonstration

Make a solution of 0.2 g of cobalt chloride-6-water in 5 cm³ of deionised water.
   Make a solution of 5 g of Rochelle salt in 60 cm³ of deionised water in a 250 cm³ beaker.

THE ROYAL
SOCIETY OF
CHEMISTRY

The demonstration

Add 20 cm³ of 20 volume hydrogen peroxide to the solution of Rochelle salt and heat the mixture to about 75 °C over a Bunsen burner. There will be a slow evolution of gas showing that the reaction is proceeding. Stirring the solution makes the evolution of gas more obvious. Now add the cobalt chloride solution to the mixture. Almost immediately the pink solution will turn green and after a few seconds vigorous evolution of gas starts and the froth will rise almost to the top of the beaker. Within about 30 seconds, the frothing subsides and the pink colour returns.

Visual tips

Stand the beaker on a overhead projector to make the evolution of gas before the addition of the catalyst more easily visible.

Teaching tips

The green activated complex can be trapped if a sample of the green solution is withdrawn with a dropping pipette and then transferred to a test-tube that is cooled in crushed ice. The solution remains green for some time.

If the reaction is considered to be going too fast for easy observation, carry it out at a lower temperature (although this will make it less easy to see the evolution of CO_2 before adding the catalyst).

Theory

The basic reaction appears to be:

$$5H_2O_2(aq) + C_4H_4O_6{}^{2-}(aq) \rightarrow 4CO_2(g) + 2OH^-(aq) + 6H_2O(l)$$

The equation may also be written in two parts:

$$3H_2O_2(aq) + C_4H_4O_6{}^{2-}(aq) \rightarrow 2CO_2(g) + 2HCOO^-(aq) + 4H_2O(l)$$

$$2HCOO^-(aq) + 2H_2O_2(aq) \rightarrow 2CO_2(g) + 2H_2O(l) + 2OH^-(aq)$$

The reaction is catalysed by pink Co^{2+} ions which are first oxidised to green Co^{3+} ions (complexed by tartrate ions) and then reduced back to Co^{2+}.

While the majority of the gas evolved is carbon dioxide, oxygen will also be produced from the decomposition of some of the hydrogen peroxide. The gas mixture will turn limewater milky, but does not extinguish a glowing splint.

Extensions

Cobalt(II) bromide also catalyses the reaction and students could be asked to try other cobalt salts. The reaction is easy to time and could form the basis of an investigation into the factors affecting reaction rates.

Further details

There are more details of the mechanism in Inner London Education Authority, *Independent Learning Project for Advanced Chemistry (ILPAC)*, Unit I5, p 57– 61. London: John Murray, 1984. For note, however, that the procedure given in this book for the experiment does not appear to work satisfactorily.

Safety

Wear eye protection.

Take care when placing the solution on the OHP.

It is the responsibility of teachers doing this demonstration to carry out an appropriate risk assessment.

2. An oscillating reaction

Topic

Reaction rates. Oscillating reactions are not part of school or college syllabuses, but are spectacular and worth doing to stimulate student interest in chemistry. They are also popular at open days and other public demonstrations.

Timing

About 10 min.

Level

Any, if demonstrated for interest, but the experiment could be used at post-16 level to introduce a topic on reaction rates and mechanisms.

Description

This is one of the simplest oscillating reactions to demonstrate. Bromate ions oxidise malonic acid to carbon dioxide. The reaction is catalysed by manganese(II) ions. On mixing the reactants and catalyst, the reaction mixture oscillates in colour between red-brown (bromine, an intermediate) and colourless with a time period of about 20 seconds.

Apparatus

▼   Magnetic stirrer and follower (optional).

▼   1 dm³ beaker.

Chemicals

The quantities given are for one demonstration.

▼   75 cm³ of **concentrated sulphuric acid** (H_2SO_4).

▼   9 g of **malonic acid** (propanedioic acid, $CH_2(CO_2H)_2$).

▼   8 g of **potassium bromate(V)** ($KBrO_3$).

▼   1.8 g of manganese(II) sulphate ($MnSO_4.H_2O$).

▼   750 cm³ of deionised water.

Method

Before the demonstration

Place 750 cm³ deionised water in a 1 dm³ beaker. Slowly, and with stirring, add 75 cm³ concentrated sulphuric acid carefully. The mixture will heat up to about 50 °C. Allow it to cool back to room temperature. This will take some time. Weigh out the malonic acid, potassium bromate and manganese sulphate in weighing boats.

The demonstration

Place the beaker of sulphuric acid on a magnetic stirrer and stir the solution fast enough for a vortex to form. A stirring rod can be used, but is tedious and tends to detract from the demonstration. Add the malonic acid and potassium bromate. When these have dissolved, add the manganese sulphate and observe what happens. A red

THE ROYAL
SOCIETY OF
CHEMISTRY

colour should develop immediately. This will disappear after about one minute and thereafter the colour will oscillate from red to colourless with a time period of about 20 seconds for a complete oscillation. This will continue with a gradually increasing time period for over ten minutes – long enough for most audiences to lose interest!

Visual tips

A white background is useful.

Teaching tips

A member of the audience with a stopwatch could be asked to time the oscillation.

Theory

This reaction is an example of a class of reactions called Belousov-Zhabotinsky (BZ) reactions. The overall reaction is usually given as:

$$3CH_2(CO_2H)_2(aq) + 4BrO_3^-(aq) \rightarrow 4Br^-(aq) + 9CO_2(g) + 6H_2O(l)$$

Oscillation is brought about by two autocatalytic steps. Bromine is an intermediate in the reaction scheme – the red colour that is observed. An analogy with predator-prey relationships might be one way to give students some idea of what is going on. For example a population of rabbits (analogous to the bromine) will increase rapidly (exponentially) if there is plenty of food (reactants). However, the plentiful supply of rabbits will stimulate a rapid increase in the fox population (another intermediate that reacts with the bromine) which will then deplete the rabbits. Lacking rabbits, the foxes will die, bringing us back to square one, ready for a rapid increase in rabbits and so on.

Extensions

The reaction can be investigated using a colorimeter with a chart recorder or interfaced to a computer. See for example R. Edwards, *Interfacing Chemistry Experiments*, London: RSC, 1993.

Further details

The reaction will not work if tap water (Coventry) is used instead of deionised water. Chloride ions, via the addition of a pinch of potassium chloride or dilute hydrochloric acid will immediately stop the oscillations. Clean apparatus is therefore essential.

The theory of oscillating reactions is complex and not fully understood. Some of the more accessible articles are listed below.
D. O. Cooke, *Educ. Chem.*,1975, **12**, 144.
I. R. Epstein, *Chem. Eng. News.*, 1987, **65**, 24.
I. R. Epstein *et al, Sci. Amer.*, 1983, **248**, 112.
M. D. Hawkins *et al, Educ. Chem.*, 1977, **14**, 53.

Safety

Wear eye protection.

The reaction mixture can be washed down the sink with plenty of water after the demonstration.

It is the responsibility of teachers doing this demonstration to carry out an appropriate risk assessment.

THE ROYAL
SOCIETY OF
CHEMISTRY

3. Extracting iron from breakfast cereal

Topic

Food science, transition metals, general interest.

Timing

About 10 min.

Level

Any.

Description

A magnetic stirrer is used to extract some of the added iron from breakfast cereal.

Apparatus

▼   One 1 dm³ beaker;

▼   magnetic stirrer and follower (ideally one coated in white Teflon);

▼   Tweezers.

Chemicals

Breakfast cereal with added iron such as 'Special K'.

Method

The demonstration
Measure about 50 g (or one serving) of cereal into the beaker. Crush the cereal by hand and add about 500 cm³ of water. Stir the mixture using the magnetic stirrer and follower for a few minutes. Remove the follower using the tweezers. A small but noticeable amount of iron powder will be seen sticking to it.

Visual tips

The amount of iron is small so it will be necessary to pass the follower around the audience, in a plastic weighing boat for example.

Theory

Manufacturers add iron to some cereals – and other food products such as flour – as a finely divided powder. This will dissolve in stomach acid before being absorbed by the body. It is added in this form (before cooking) because it does not produce any taste or interact chemically with other components of the product. 'Special K' packets quote 20 mg of iron per 100 g of cereal. Products such as cornflakes which are fortified at a lower level have about 6.7 mg of iron per 100 g while un-fortified breakfast cereals have 1–2 mg of iron per 100 g.

Extensions

Repeat the demonstration with other cereals or other foods.

Safety

Wear eye protection.
    The slurry resulting from the demonstration can be safely poured down the sink. It is probably worth re-stating the rules about not eating in laboratories.
    It is the responsibility of teachers doing this demonstration to carry out an appropriate risk assessment.

THE ROYAL
SOCIETY OF
CHEMISTRY

Acknowledgement

This procedure is adapted from an idea by David Katz of the Community College,
Philadelphia, US.

THE ROYAL
SOCIETY OF
CHEMISTRY

4. The equilibrium between ICl and ICl$_3$

Topic

The effect of reactant concentration and reaction temperature on equilibrium position.

Timing

About 10 min.

Level

Post-16 and possibly pre-16.

Description

The equilibrium $ICl(l) + Cl_2(g) \rightleftharpoons ICl_3(s)$ is set up in a U-tube and the concentration of chlorine can be varied to demonstrate its effect on the position of the equilibrium. The effect of temperature can also be shown clearly. Excess chlorine is absorbed by sodium hydroxide.

Apparatus

▼ Two 250 cm^3 conical flasks with two-hole stoppers.

▼ U-tube with stoppers.

▼ Dropping funnel.

▼ Three-way tap.

▼ Filter pump.

▼ Two screw clips.

▼ Two 2 dm^3 beakers.

▼ Glass tubing and plastic tubing (*see figure*).

▼ Access to a fume cupboard.

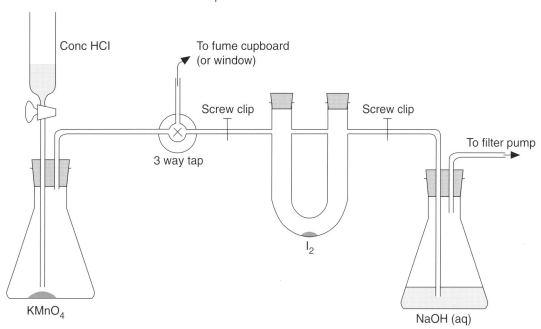

Equilibrium between ICl and ICl$_3$

THE ROYAL
SOCIETY OF
CHEMISTRY

Chemicals

The quantities given are for one demonstration.

▼   Approximately 100 cm³ of **concentrated hydrochloric acid**.

▼   Approximately 10 g of **potassium permanganate** (potassium manganate(VII), $KMnO_4$).

▼   Approximately 0.1 g of **iodine**.

▼   Approximately 100 cm³ of **sodium hydroxide** solution of concentration at least 2 mol dm⁻³

▼   About 250 cm³ of crushed ice.

Method

Before the demonstration

Set up the apparatus as shown in the figure with about 0.1 g of iodine in the U-tube. Clear plastic tubing can be used for the connections – pressure tubing is not necessary. Run a length of tubing from the three-way tap to a fume cupboard (switched on!) or through an outside window. If doing the latter, check that this will not present any hazard. The apparatus can be used in the open laboratory.

The demonstration

Turn the three-way tap so that the left hand arm of the U-tube draws air through the apparatus and turn on the filter pump to give a gentle bubbling through the sodium hydroxide. Now turn the three way tap to connect the U-tube to the **chlorine** generator and drip a little hydrochloric acid onto the potassium permanganate. As the chlorine is drawn over the iodine, a brown liquid (iodine monochloride, ICl) is first formed. Manipulate the dropping funnel to maintain the supply of chlorine. Within a few seconds a yellow solid (iodine trichloride, ICl_3) will form inside the right hand arm of the U-tube. Now turn the three-way tap to connect the chlorine generator to the outside (to avoid a build-up of pressure in the generator). Simultaneously remove the stopper from the left hand arm of the U-tube so that air is drawn through the apparatus, displacing the chlorine in the U-tube. The yellow solid will turn back to a brown liquid. Replacing the stopper on the U-tube and turning the three-way tap to restore the chlorine supply will result in the yellow solid re-appearing. The demonstration can be repeated several times. Note that the first appearance of the iodine trichloride is the most clearly seen, probably because it sticks better to the clean glass.

To demonstrate the effect of temperature on the equilibrium, turn the three-way tap to connect the chlorine generator to the outside to prevent pressure build-up and tighten the screw clips at the two positions shown in the diagram to isolate the U-tube. This can now be removed from the rest of the apparatus. Dip the bottom of the U-tube into a beaker of water that has just boiled. The contents of the tube will turn to the brown liquid ICl. Now place the bottom of the U-tube into a beaker of ice water. The contents of the beaker will form the yellow solid ICl_3. This cycle can be repeated many times.

Visual tips

Arrange a suitable background so that the audience can see the changes in the U-tube clearly. Direct the audience's attention to the U-tube so that they are not distracted by the rest of the relatively complex apparatus.

THE ROYAL
SOCIETY OF
CHEMISTRY

Teaching tips

Students will probably not be familiar with interhalogen compounds but should be able to see that they are perfectly feasible in bonding terms. Point out that it is reasonable that ICl is a liquid in view of its lower relative molecular mass than that of iodine and similarly that ICl_3 should be a solid.

Theory

The equilibrium

$$ICl(l) \; + \; Cl_2(g) \; \rightleftharpoons \; ICl_3(s) \quad \Delta H = -105.9 \text{ kJ mol}^{-1}$$
brown   green   yellow

is set up and the demonstration shows that it obeys Le Chatelier's principle, *ie* that increasing the chlorine concentration (or partial pressure) displaces the equilibrium to the right and that decreasing the chlorine concentration moves it to the left. Increasing the temperature displaces the equilibrium to the left and decreasing the temperature displaces it to the right.

Further details

Some teachers may prefer to use an alternative method for generating chlorine by dripping 5 mol dm^{-3} hydrochloric acid onto sodium chlorate(I) solution. However, since dry chlorine is not necessary, there is no need for sulphuric acid and therefore there is no possibility of confusing this with the hydrochloric acid, which seems to be the objection sometimes raised over the method suggested here. A chlorine cyclinder could be used if available.

Safety

Wear eye protection.

The residue in the U-tube is corrosive. It is water-soluble and can be flushed down the sink with plenty of water.

The contents of the sodium hydroxide flask can be flushed down the sink with plenty of water.

The contents of the chlorine generator can be poured down the sink in the fume cupboard with plenty of water. Do not dispose of this solution down the same sink immediately after disposing of the contents of the sodium hydroxide flask because chlorine could be produced.

It is the responsibility of teachers doing this demonstration to carry out an appropriate risk assessment.

THE ROYAL
SOCIETY OF
CHEMISTRY

5. The combustion of iron wool

Topic

Chemical reactions – increase in mass on chemical combination.

Timing

About 5 min.

Level

Introductory chemistry.

Description

Iron wool is heated on a simple 'seesaw' balance and the increase in mass is shown.

Apparatus

▼   Bunsen burner.

▼   Heat-proof mat.

▼   Metre rule.

▼   A piece of aluminium cooking foil about 10 cm x 10 cm.

▼   Retort stand, boss and clamp.

▼   A few grams of plasticine.

Chemicals

The quantities given are for one demonstration.

▼   About 4 g of steel wool.

Method

Before the demonstration

Cover the end 10 cm of a metre rule with aluminium foil to protect the wood from the Bunsen flame. Take about 4 g of steel wool, tease it out to allow air to get to it and attach the wool to the top of the metre rule at the end covered by the foil. A few strands of the iron wool will attach it well enough. Balance the rule on a suitable knife edge at the 50 cm mark (*see Fig*). A groove cut across the width of the rule with a saw helps. Weight the empty end of the rule with plasticine until this end is *just* down. This is critical. Place a heatproof mat under the iron wool.

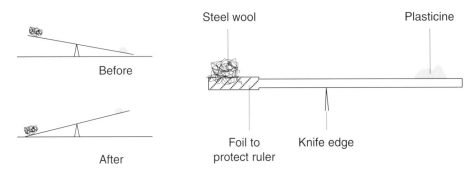

Combustion of iron wool

THE ROYAL
SOCIETY OF
CHEMISTRY

The demonstration

Heat the iron wool directly with a roaring Bunsen flame. It will glow and some glowing pieces of wool will drop off onto the heatproof mat. Heat for about a minute by which time the metre rule will have overbalanced so that the iron wool side is down.

Teaching tips

While heating, ask the students to predict whether the weight of the iron wool will go up, down or remain the same. Many will predict a weight loss.

Safety

Wear eye protection.

It is the responsibility of teachers doing this demonstration to carry out an appropriate risk assessment.

THE ROYAL
SOCIETY OF
CHEMISTRY

6. Chemiluminescence – oxidation of luminol

Topic

Energy changes; this demonstration shows that the energy of a chemical reaction can be given out as light as well as heat. However the demonstration could be performed simply to stimulate interest in chemistry or at an open day or other public demonstrations.

Timing

Less than 5 min.

Level

14–18 age group for teaching purposes; any age for general interest.

Description

An aqueous solution of luminol (3-aminophthalhydrazide) is oxidised by a solution of sodium chlorate(I) (commercial bleach) and gives out a blue glow without increase in temperature.

Apparatus

▼   Two 1 dm³ conical flasks with stoppers.

▼   One 2 dm³ beaker.

▼   0 – 100 °C thermometer.

Chemicals

The quantities given are for one demonstration.

▼   0.4 g of luminol (3-aminophthalhydrazide).

▼   4.0 g of **sodium hydroxide.**

▼   100 cm³ of household **bleach** (approximately 5 % NaOCl).

Method

Before the demonstration

Mix 100 cm³ of bleach and 900 cm³ of water in one of the flasks and stopper the flask. In the other flask put 0.4 g of luminol, 1 dm³ of water and 4 g of sodium hydroxide. Swirl to dissolve the chemicals and then put on the stopper. Tap water can be used for making up the solutions. The luminol does not appear to dissolve completely, leaving a fine greenish suspension.

The demonstration

Lower the room lights and pour the two solutions at the same rate into the beaker so that they mix on pouring. A pale blue glow will be seen on mixing which persists for a few seconds. Take the temperatures of the solutions and the resulting mixture. They will be the same.

THE ROYAL
SOCIETY OF
CHEMISTRY

Visual tips

The darker the room the better, provided that the demonstrator can see to pour! The temperature readings will be better seen if a thermocouple-type thermometer with a large display or one interfaced to a computer with a display on a monitor is available.

Teaching tips

The solutions are stable for over 12 hours and so can be made up well before the demonstration.

Theory

Luminol is oxidised to the aminophthalate ion which is produced in an excited state and emits light on dropping to the ground state.

aminophthalate
ion

Extensions

The solutions can be poured into a funnel attached to a clear plastic tube which can be bent into a variety of shapes. This can enhance the visibility of the demonstration. Chemiluminescent 'light sticks' can be bought in outdoor equipment shops. These contain a glass phial containing one solution inside a plastic tube containing the other. The reaction is started by breaking the glass tube and the glow continues for some hours.

Safety

Wear eye protection.

After the experiment, the mixture can be flushed down the sink with plenty of water.

It is the responsibility of teachers doing this demonstration to carry out an appropriate risk assessment.

THE ROYAL
SOCIETY OF
CHEMISTRY

7. Determining relative molecular masses of gases

Topic

Gases and the gas laws; relative molecular mass determination.

Timing

About 5 min.

Level

Pre-16 or post-16

Description

These are simple, quick and direct methods for finding relative molecular masses of gases. They are conceptually simple and the calculations involved are easy. Gases from pressurised containers are collected and the decrease in mass of the containers is determined to allow the relative molecular masses to be calculated.

Apparatus

▼   One 1 dm³ measuring cylinder.

▼   A trough.

▼   A flexible delivery tube.

▼   Access to a top pan balance.

▼   0 – 100 °C thermometer (optional).

▼   Access to a barometer (optional).

▼   Access to a fume cupboard (optional).

Chemicals

Can of butane for re-filling lighters, or another source of gas (*see Fig*) or cigarette lighter.

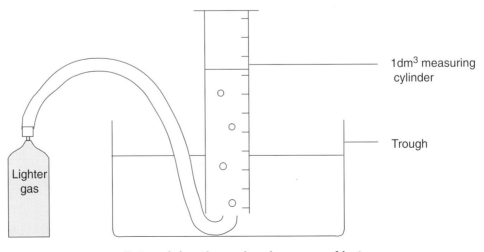

Lighter gas

1dm³ measuring cylinder

Trough

Determining the molecular mass of butane

THE ROYAL
SOCIETY OF
CHEMISTRY

Method

Before the demonstration
Attach a length of flexible tubing to one of the adapters supplied with the can of gas. Set up the trough and clamp (or get an assistant to hold) an inverted measuring cylinder full of water for gas collection over water.

The demonstration
Weigh the gas can and then collect 1 dm³ of gas in the measuring cylinder. Ensure that the levels of water inside and outside the cylinder are the same after gas collection. Re-weigh the gas can making sure it is dry if hands are wet. About 2 g of gas will have been used. Take the temperature of the room and the atmospheric pressure if required.

Teaching tips

The can will feel cold to the touch. This could be pointed out to the class and the reason discussed.

Theory

Calculate the mass of one mole of the gas by one of the following methods, depending on the ability of the class.

a)   Use the approximate relationship that one mole of gas has a volume of 24 dm³ at room conditions. This calculation requires no more than the ability to multiply by 24.

b)   Use $PV = nRT$ to calculate the number of moles in the known mass of gas. In this case it will be necessary to measure the temperature and atmospheric pressure.

Extensions

Other sources of gas could be camping gas stoves and gas blowlamps. All that is required is a little ingenuity in attaching a delivery tube. In some blowlamps, for example, it is possible to remove the burner by undoing one screw.

Another possibility is to use a cigarette lighter as the source of gas. No delivery tube is needed; simply hold the lighter under water in the trough below the measuring cylinder and open the valve. The flow of gas is slow. It may take several minutes to obtain 1 dm³ and it may be better to collect, say, 250 cm³ – which still gives an easily measurable weight loss (about 0.4 g). It is possible to use a rubber band to hold the valve open. The lighter must be dried carefully before re-weighing. Wiping with absorbent soft tissue seems to work satisfactorily, but some people might want to use a hairdrier as well.

It is also possible to adapt the method to collect and investigate the propellant gases from aerosol containers, see reference below.

Further details

J. M. Lister, *Practical work with aerosols and pressurised gases*, *Sch. Sci. Rev.*, 1987, 516.

Safety

Wear eye protection.

The gases used are flammable; make sure that there are no sources of ignition nearby. Empty the cylinder of gas outside or in the fume cupboard.

It is the responsibility of teachers doing this demonstration to carry out an appropriate risk assessment.

THE ROYAL
SOCIETY OF
CHEMISTRY

8. The equilibrium between $Co(H_2O)_6^{2+}$ and $CoCl_4^{2-}$

Topic

Equilibrium; factors that affect the equilibrium position.

Timing

About 5 min.

Level

Post-16.

Description

$Co(H_2O)_6^{2+}$ is pink in aqueous solution and $CoCl_4^{2-}$ is blue. The equilibrium between these two species can be disturbed by adding Cl^- ions and by changing the temperature. The changes in the equilibrium position are as predicted by Le Chatelier's principle.

Apparatus

▼ Six boiling tubes and a rack.

▼ One 100 cm^3 measuring cylinder.

▼ Three 250 cm^3 beakers.

▼ Two dropping pipettes.

▼ Access to a top pan balance.

Chemicals

The quantities given are for one demonstration.

▼ 4 g **cobalt(II) chloride-6-water** ($CoCl_2.6H_2O$).

▼ 100 cm^3 of **concentrated hydrochloric acid**.

▼ About 200 cm^3 of crushed ice.

Method

Before the demonstration
Boil a beaker of water and prepare a beaker of crushed ice and water.
Dissolve about 4 g of cobalt(II) chloride-6-water in 40 cm^3 of water. A pink solution containing $Co(H_2O)_6^{2+}$ will be formed.

The demonstration
Take the pink cobalt chloride solution and make it up to 100 cm^3 with concentrated hydrochloric acid using a measuring cylinder. A violet solution will be formed. Adding more hydrochloric acid will produce a blue solution containing $CoCl_4^{2-}$, while adding water will restore the pink colour.

    By trial and error produce an 'in between' violet coloured solution containing the two cobalt ions and place about 2 cm depth of it in each of the six boiling tubes. Place these on the bench in two groups of three.

THE ROYAL
SOCIETY OF
CHEMISTRY

1. Effect of concentration

Keeping one tube as a control, add water to one tube and concentrated hydrochloric acid to the other using dropping pipettes. If desired, show that these changes are reversible.

2. Effect of temperature

Keeping one tube as a control, place another tube in the hot water (over 90 °C). It will go blue. Put the third tube in the ice/water mixture. It will go pink. If desired, show that the changes are reversible.

Visual tips

A white background will help. For a large audience the effect is clearer if the reactions are scaled up and done in beakers.

Theory

The equilibrium is:

$$Co(H_2O)_6^{2+}(aq) + 4Cl^-(aq) \rightleftharpoons CoCl_4^{2-}(aq) + 6H_2O(l) \quad \Delta H \text{ +ve}$$
$$\text{pink} \qquad\qquad\qquad\qquad \text{blue}$$

The shifts in equilibrium position produced by both temperature and concentration changes are in accordance with Le Chatelier's principle.

Extensions

It is possible to show that it is the Cl^- ions in the hydrochloric acid that shift the equilibrium by adding a spatula of sodium chloride to the pink solution. This produces a bluer colour eventually although this takes some time to form because the salt is slow to dissolve.

Safety

Wear eye protection.

It is the responsibility of teachers doing this demonstration to carry out an appropriate risk assessment.

THE ROYAL
SOCIETY OF
CHEMISTRY

9. Phenolphthalein as an indicator

Topic

Indicators, acids and bases, equilibria.

Timing

About 5 min.

Level

Post-16.

Description

As well as the well-known 'red in alkali, colourless in acid', phenolphthalein will go colourless in more concentrated alkaline solutions.

Apparatus

▼   One 100 cm³ beaker.

▼   Three boiling tubes and rack.

▼   Dropping pipette.

Chemicals

The quantities given are for one demonstration.

▼   Phenolphthalein indicator solution (0.1 g solid dissolved in 60 cm³ of **ethanol** and 40 cm³ of water).

▼   A few pellets of solid **sodium hydroxide**.

▼   About 100 cm³ of 0.5 mol dm⁻³ **sodium hydroxide**.

▼   A few cm³ of approximately 2 mol dm⁻³ **hydrochloric acid**.

Method

The demonstration
Add a few drops of phenolphthalein solution to about 100 cm³ of 0.5 mol dm⁻³ sodium hydroxide in a beaker until a deep pink colour is visible. Divide this solution between the three boiling tubes. Leaving one tube as a control, add hydrochloric acid dropwise to one of the other two until the colour disappears. This is the 'usual' behaviour of phenolphthalein. To the third tube, add two or three pellets of solid sodium hydroxide and swirl to dissolve. The pink colour will disappear in this tube too. The colour changes can be reversed by appropriate additions of acid or alkali.

Visual tips

A white background will help.

Teaching tips

The demonstration could be presented by starting with the colourless solution of phenolphthalein in concentrated alkali and adding acid to it to give an unexpected colour change.

Theory

In acid solution, phenolphthalein occurs as (I) which is colourless. Addition of sodium hydroxide removes two protons to produce the red dianion (II), while further addition of alkali gives (III) which is colourless.

(I)
colourless

$2\,OH^-$ ⇌ $2\,H^+$

(II)
red

OH^- ⇌ H^+

(III)
colourless

Safety

Wear eye protection.

It is the responsibility of teachers doing this demonstration to carry out an appropriate risk assessment.

THE ROYAL
SOCIETY OF
CHEMISTRY

10. Strong and weak acids: the common ion effect

Topic

Strong acids and weak acid equilibria.

Timing

Between 5 and 15 min depending on how much is done in front of the class.

Level

Post-16.

Description

The rates and extent of the reactions of hydrochloric acid, ethanoic acid, and a mixture of ethanoic acid and sodium ethanoate with calcium carbonate are compared approximately by looking at the heights of froth produced.

Apparatus

▼   Three 250 cm^3 measuring cylinders.

▼   Three 250 cm^3 conical flasks (with approximate graduations every 100 cm^3).

▼   Two dropping pipettes.

▼   Access to a top pan balance.

Chemicals

The quantities given are for one demonstration.

▼   100 cm^3 of 2 mol dm^{-3} **hydrochloric acid**.

▼   100 cm^3 of 2 mol dm^{-3} ethanoic acid (acetic acid).

▼   3 g of calcium carbonate powder (precipitated).

▼   13.5 g of sodium ethanoate -3-water (sodium acetate trihydrate, $CH_3CO_2Na.3H_2O$).

▼   A few cm^3 of Teepol or other detergent.

▼   A few cm^3 of universal indicator solution.

Method

Before the demonstration

Prepare the two acid solutions if necessary. Hydrochloric acid: add 170 cm^3 of **concentrated (36 %) hydrochloric acid** to 600 cm^3 of deionised water and make up the resulting solution to 1 dm^3. Ethanoic acid: add 115 cm^3 of **glacial ethanoic acid** to 600 cm^3 of deionised water and make up the resulting solution to 1 dm^3.

Weigh 1 g of calcium carbonate powder into each of three 250 cm^3 measuring cylinders. Add about 1 cm^3 of Teepol or washing up liquid to each. Label them HCl, CH_3CO_2H and $CH_3CO_2H + CH_3CO_2Na$ and line them up on the bench.

Weigh 13.5 g of sodium ethanoate-3-water into a conical flask. Dissolve this in about 80 cm^3 of 2 mol dm^{-3} ethanoic acid and make the resulting mixture up to 100 cm^3 with more ethanoic acid solution (the graduations on the flask will be enough).

THE ROYAL
SOCIETY OF
CHEMISTRY

This solution is approximately 1 mol dm^{-3} with respect to sodium ethanoate and approximately 2 mol dm^{-3} with respect to ethanoic acid.

The demonstration

Pour 100 cm^3 of 2 mol dm^{-3} ethanoic acid into one of the remaining conical flasks and 100 cm^3 of 2 mol dm^{-3} hydrochloric acid into the other. Add enough universal indicator solution to each flask to give an easily visible colour. This will be red in the hydrochloric acid, orange in the ethanoic acid and yellow in the mixture of ethanoic acid and sodium ethanoate.

With the help of an assistant, pour the contents of each conical flask simultaneously into the appropriate measuring cylinder. The hydrochloric acid will froth fastest, producing about 220 cm^3 of froth in a few seconds. The ethanoic acid will be a little slower, but will eventually produce about the same height of froth. The mixture will be much slower and will produce only about half the height of froth of the other two.

Visual tips

A black background may be better than a white one for viewing the froth.

Teaching tips

The ethanoic acid/sodium ethanoate mixture produces less froth than the other two because the foam is not stable for more than a few seconds and not because the total acidity is lower. This may need to be pointed out to the students. Point out that all of the calcium carbonate reacts eventually. Students could be asked to calculate the number of moles of acid in each solution and the number of moles of calcium carbonate to show that the former is in excess. They could then calculate the volume of gas expected (approximately 240 cm^3) and confirm that the amount of froth is reasonable. If they have done sufficient theory, they could be asked to calculate the pH of each solution for homework.

Theory

Hydrochloric acid is a strong acid and is dissociated completely into H$^+$(aq) and to Cl$^-$(aq) in aqueous solution. Hence the initial rate of reaction with the carbonate is high. Ethanoic acid is a weak acid (K$_a$ = 1.7 x 10^{-5} mol dm^{-3}) so the concentration of H$^+$(aq) is much lower. In the mixture of sodium ethanoate and ethanoic acid, the presence of the ethanoate ion from the sodium ethanoate pushes the equilibrium

$$CH_3CO_2H(aq) \rightleftharpoons CH_3CO_2^-(aq) + H^+(aq)$$

to the left, reducing the H$^+$(aq) concentration still further. However, for both the ethanoic acid and the ethanoic acid /sodium ethanoate mixture, as hydrogen ions are used up in the reaction, more ethanoic acid dissociates hence the total amount of H$^+$(aq) is the same in each case.

Extensions

A pH meter could be used to measure the pH of the three acid solutions before the reaction.

Safety

Wear eye protection.

It is the responsibility of teachers doing this demonstration to carry out an appropriate risk assessment.

THE ROYAL
SOCIETY OF
CHEMISTRY

11. The reaction between zinc and copper oxide

Topic

Chemical reactions (displacement, exothermic).

Timing

About 5 min.

Level

Introductory chemistry.

Description

Copper oxide and zinc are heated together. The reaction is exothermic and the products can be identified clearly.

Apparatus

▼   Bunsen burner and heatproof mat.

▼   Tin lid.

▼   One 100 cm³ beaker.

▼   Circuit board with batteries, bulb and leads (optional).

▼   Safety screen.

Chemicals

The quantities given are for one demonstration.

▼   4 g of **copper(II) oxide** powder.

▼   1.6 g of zinc powder.

▼   Approximately 20 cm³ of dilute (approximately 2 mol dm⁻³) **hydrochloric acid**.

▼   A few grams each of zinc oxide and copper powder.

Method

The demonstration

Weigh out 2 g (0.025 mol) of copper oxide and 1.6 g (0.025 mol) of zinc powder and mix thoroughly to give a uniformly grey powder. Pour the mixture in the shape of a 'sausage' about 5 cm long onto a heat-proof mat or clean tin lid. Heat one end of the 'sausage' from above with a roaring Bunsen flame until it begins to glow, then remove the flame. A glow will spread along the 'sausage' until it has all reacted. A white/grey mixture will remain. Heat this to show that the white powder (zinc oxide) is yellow when hot and white when cool. Pour the cool residue into a 100 cm³ beaker and add a little dilute hydrochloric acid to dissolve the zinc oxide (and also any unreacted zinc and copper oxide) with warming if necessary. Red-brown copper will be left. This can be rinsed with water and passed around the class for identification. Show that the powder conducts electricity using a circuit board. If further confirmation of identity is required, heat the red-brown powder with concentrated nitric acid to give a blue solution of copper nitrate.

THE ROYAL
SOCIETY OF
CHEMISTRY

Visual tips

Scale up the quantities if the audience is some distance away.

Teaching tips

Demonstrate that zinc oxide goes yellow when heated and returns to white when cool to help confirm the identity of this product. (This phenomenon is caused by a change in crystal structure.)

Extensions

Other metals can be used, but take care to compare like with like. Coarse magnesium powder, for example, gives a less vigorous reaction than powdered zinc. Finely powdered magnesium gives a very vigorous reaction.

Safety

Wear eye protection.

It may be considered necessary to place a safety screen between the experiment and the audience.

It is the responsibility of teachers doing this demonstration to carry out an appropriate risk assessment.

THE ROYAL
SOCIETY OF
CHEMISTRY

12. Catalysis of the reaction between sodium thiosulphate and hydrogen peroxide

Topic

Reaction rates and catalysis.

Timing

About 10 min.

Level

Introductory reaction rates and catalysis.

Description

Hydrogen peroxide oxidises sodium thiosulphate to sulphuric acid. Starting from an alkaline solution, the resulting pH change is followed using universal indicator which changes from blue to green to yellow to orange-red. Adding an ammonium molybdate catalyst speeds up the colour changes considerably.

Apparatus

▼   Four 1 dm³ flasks.

▼   One 100 cm³ measuring cylinder.

▼   One 100 cm³ beaker.

Chemicals

The quantities given are for one demonstration.

▼   8.7 g of sodium thiosulphate-5-water.

▼   3.8 g of sodium ethanoate-3-water (sodium acetate tri-hydrate), or 2.3 g of anhydrous sodium ethanoate.

▼   0.5 g of **sodium hydroxide**.

▼   0.08 g of ammonium molybdate(VI).

▼   14 cm³ of **100 volume hydrogen peroxide**.

▼   A few cm³ of universal indicator solution.

▼   1.1 dm³ of deionised water.

Method

Before the demonstration

Dissolve the above masses of sodium thiosulphate, sodium ethanoate and sodium hydroxide together in deionised water and make up to 1 dm³. Add sufficient universal indicator solution to give an easily visible blue colour. Pour 225 cm³ of this solution into each of three 1 dm³ flasks labelled 'catalyst', 'no catalyst' and 'control', respectively. Make a solution of 14 cm³ of 100 volume hydrogen peroxide made up to 40 cm³ with deionised water. Divide this into two 20 cm³ portions. Weigh out 0.08 g of ammonium molybdate.

THE ROYAL
SOCIETY OF
CHEMISTRY

The demonstration

Place the three flasks containing the blue solution on the bench. Add the weighed sodium molybdate to the one labelled 'catalyst' and swirl to dissolve it. Then, at the same time, add the 20 cm³ portions of hydrogen peroxide solution to the flasks marked 'catalyst' and 'no catalyst', leaving the third flask as a control to aid colour comparison. Over three or four minutes the solution with the catalyst will change from blue through green, yellow and orange to orange-red. The solution without the catalyst will follow the same sequence but more slowly, although it will not have reached the same red-orange colour of the first solution after an hour.

Visual tips

A white background helps the audience to see the colour changes clearly.

Theory

The reaction is:

$$Na_2S_2O_3(aq) + 4H_2O_2(aq) \rightarrow Na_2SO_4(aq) + H_2SO_4(aq) + 3H_2O(l)$$

The sulphuric acid produced by the reaction neutralises the sodium hydroxide (buffered by the sodium ethanoate) and brings about the observed colour changes.

Further details

One teacher involved in the trial reports carrying out the reaction with 20 volume hydrogen peroxide rather than diluted 100 volume peroxide. The reaction was a little slower.

Safety

Wear eye protection.

It is the responsibility of teachers doing this demonstration to carry out an appropriate risk assessment.

THE ROYAL
SOCIETY OF
CHEMISTRY

13. The optical activity of sucrose

Topic

Optical activity/optical isomerism.

Timing

About 5 min.

Level

Post-16.

Description

The ability of a concentrated sucrose (table sugar) solution to rotate the plane of polarisation of polarised light is shown.

Apparatus

▼   A pair of polaroid sunglasses (or two polaroid filters).

▼   One 400 cm³ beaker.

▼   Overhead projector.

Chemicals

The quantity given is for one demonstration.

▼   About 250 g of *D*-sucrose (table sugar).

Method

Before the demonstration
Dissolve about 250 g of sugar in about 250 cm³ of warm water. This will take some time. Decant the syrupy solution off any remaining solid sugar and transfer it to a 400 cm³ beaker. Remove one of the lenses from a pair of polaroid sunglasses (this can usually be replaced quite easily).

The demonstration
Place the polaroid lens in the centre of the stage of an overhead projector (OHP). Using the frame of the glasses as a handle, rotate the second lens above the first until a position of minimum transmitted light is found *(see Fig)*. Mark on the OHP where this position is. The second lens will be at right angles to the first. Now support the beaker of sugar solution above the first polaroid – a couple of thin strips of wood will do. Hold the second polaroid above the beaker and again rotate it to find the minimum light position and mark this on the OHP. This will be about 30⁰ clockwise from the first position.

Visual tips

With practice, the frame of the glasses can be used as a handle, and as a pointer to mark the position of the lens.

Teaching tips

It may be worth showing that pure water has no effect confirming that it is the sugar that is optically active.

A more professional apparatus can be produced using polaroid filters and a protractor, but the appeal of the experiment as described is demonstrating the effect using simple everyday items.

Theory

D-sucrose rotates the plane of polarisation clockwise as seen by an observer looking along the light beam towards the source. This clockwise rotation is denoted (+).

Extensions

It is possible to show semi-quantitatively that the rotation depends on sample concentration and optical path length.

Safety

Wear eye protection.
It is the responsibility of teachers doing this demonstration to carry out an appropriate risk assessment.

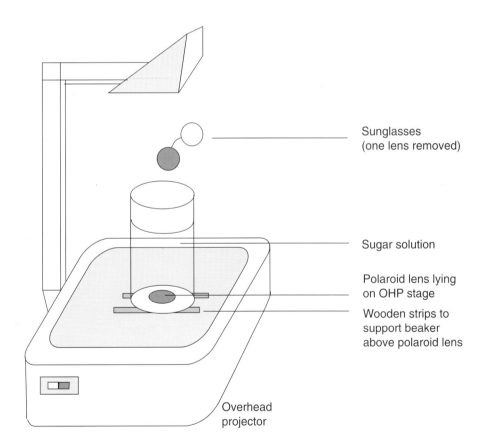

Sunglasses
(one lens removed)

Sugar solution

Polaroid lens lying
on OHP stage

Wooden strips to
support beaker
above polaroid lens

Overhead
projector

Finding the optical activity of sucrose

THE ROYAL
SOCIETY OF
CHEMISTRY

14. A sodium ethanoate stalagmite

Topic

Solutions, supersaturation, general interest.

Timing

Less than 1 min.

Level

Any.

Description

Rapid crystallisation of a supersaturated solution produces a stalagmite.

Apparatus

▼   One 250 cm³ beaker.

▼   One watch glass.

▼   Stirring rod.

▼   Bunsen burner, tripod and gauze.

▼   Access to a top pan balance.

Chemicals

The quantity given is for one demonstration.

▼   125 g of sodium ethanoate-3-water (sodium acetate-3-water, $CH_3CO_2Na.3H_2O$).

Method

Before the demonstration
Weigh 125 g of sodium ethanoate into the beaker and add 12.5 cm³ of water. Heat the beaker over the Bunsen burner and stir until a clear solution is obtained. Cover with a watch glass and allow it to cool to room temperature.

The demonstration
Place a few crystals of sodium ethanoate onto a watch glass and pour the solution over these. The supersaturated solution crystallises immediately forming a 'stalagmite'. The watch glass becomes warm.

Visual tips

A black background is probably better than white for this demonstration.

Teaching tips

The stalagmite can be re-heated and used again. Keep the solution clean and free from dust, which could cause premature crystallisation. If the re-heating is shown to the class, emphasise that the solid is dissolving, not melting.

Theory

The sodium ethanoate is dissolving largely in its own water of crystallisation.

THE ROYAL
SOCIETY OF
CHEMISTRY

Extensions

Commercial 'heat packs' are available which use the principle of supersaturation. A supersaturated solution of sodium thiosulphate is stable until a seed crystal is added when it will crystallise exothermically. The pack can be re-used by heating the thiosulphate to re-dissolve it.

Safety

Wear eye protection.

It is the responsibility of teachers doing this demonstration to carry out an appropriate risk assessment.

THE ROYAL
SOCIETY OF
CHEMISTRY

15. Urea-methanal polymerisation

Topic

Polymers (condensation).

Timing

About 5 min.

Level

Introductory polymerisation or post-16 to introduce discussion of reaction
mechanism.

Description

A hard, white thermosetting polymer is produced by adding acid to a solution of urea
in formalin.

Apparatus

▼   One 100 cm³ measuring cylinder.

▼   One throw-away container such as a screw-top coffee jar.

▼   Dropping pipette.

▼   Access to a fume cupboard.

Chemicals

The quantities given are for one demonstration.

▼   20 cm³ of **formalin** (a 37 – 40 % solution of methanal (formaldehyde, HCHO)
     in water).

▼   10 g of urea (carbamide).

▼   1 cm³ of **concentrated sulphuric acid**.

▼   Plasticine.

▼   Aluminium foil.

▼   Bunsen burner (optional).

Method

The demonstration

Work in a fume cupboard. Dissolve 10 g of urea in 20 cm³ of formalin in a throw
away container such as a 100 g screw-top coffee jar to give a clear solution. Add
about 1 cm³ of concentrated sulphuric acid a drop at a time using a dropping pipette
and stir. Within a minute, the solution begins to go milky and eventually a hard white
solid is formed which is difficult to remove from the container. A lot of heat is
evolved. Show that the polymer is hard by poking the material with a spatula. The
polymer is likely to be contaminated with unreacted starting materials so it must be
washed carefully if it is to be passed around the class. It may be better to pass
around the screw-top container with the lid on.

It is possible to make a plasticine mould of a simple shape and line it with
aluminium foil to demonstrate the manufacture of articles. Immediately after adding
the acid, pour some of the solution into the mould and allow it to polymerise.

THE ROYAL
SOCIETY OF
CHEMISTRY

If desired, a sample of the polymer (from the mould) can be held with tongs and heated in a Bunsen flame. It will char but will not melt showing that it is thermosetting.

Theory

The basic reaction is a condensation polymerisation in which water is eliminated.

$$
\begin{array}{ccc}
\underset{\text{Urea}}{
\begin{array}{c}
\text{H} \quad \text{O} \quad \text{H} \\
| \quad\quad || \quad\quad | \\
\text{H}-\text{N}-\text{C}-\text{N}-\text{H}
\end{array}}
\quad + \quad
\underset{\text{Methanal}}{
\begin{array}{c}
\text{O} \\
|| \\
\text{C} \\
\diagup \quad \diagdown \\
\text{H} \quad\quad \text{H}
\end{array}}
\quad \xrightarrow{\;H_2O\;} \quad
\underset{\text{Urea-methanal}}{
\begin{array}{c}
\text{H} \quad \text{H} \quad \text{O} \quad \text{H} \quad \text{H} \\
| \quad\; | \quad\; || \quad\; | \quad\; | \\
-\text{C}-\text{N}-\text{C}-\text{N}-\text{C}- \\
| \quad\quad\quad\quad\quad\quad\quad\quad | \\
\text{H} \quad\quad\quad\quad\quad\quad\quad \text{H}
\end{array}}
\end{array}
$$

The product has many crosslinks:

$$
\begin{array}{c}
\quad\quad\quad\quad\quad\quad \text{O} \quad\quad\quad\quad\quad\quad\quad\quad \text{O} \\
\quad\quad\quad\quad\quad\quad || \quad\quad\quad\quad\quad\quad\quad\quad || \\
-\text{H}_2\text{C}-\text{N}-\text{C}-\text{N}-\text{CH}_2-\text{N}-\text{C}-\text{N}- \\
\quad\quad\quad | \quad\quad\quad\quad\quad | \quad\quad\quad\quad\quad | \\
\quad\quad\quad \text{H} \quad\quad\quad\quad \text{H} \quad \text{O} \quad \text{CH}_2 \quad\quad \text{H} \\
\quad\quad\quad\quad\quad\quad\quad\quad\quad\quad || \quad\; | \\
\quad\quad\quad\quad\quad\quad\quad\quad -\text{N}-\text{C}-\text{N} \\
\quad\quad\quad\quad\quad\quad\quad\quad\quad | \quad\quad\quad | \\
\quad\quad\quad\quad\quad\quad\quad\quad\quad \text{H} \quad\quad \text{CH}_2 \\
\quad\quad\quad\quad\quad\quad\quad\quad\quad\quad\quad\quad\quad | \\
\quad\quad\quad\quad\quad\quad \text{H}-\text{N}-\text{C}-\text{N}-\text{CH}_2- \\
\quad\quad\quad\quad\quad\quad\quad\quad\quad\quad || \quad\; | \\
\quad\quad\quad\quad\quad\quad\quad\quad\quad\quad \text{O} \quad \text{CH}_2 \\
\quad\quad\quad\quad\quad\quad\quad\quad\quad\quad\quad\quad\quad |
\end{array}
$$

Safety

Wear eye protection and work in a fume cupboard.

If the polymer is made in a disposable screw-top container, this will aid disposal as solid waste.

It is the responsibility of teachers doing this demonstration to carry out an appropriate risk assessment.

THE ROYAL
SOCIETY OF
CHEMISTRY

16. Dalton's law of partial pressures

Topic

The gas laws.

Timing

About 10 min.

Level

Post-16.

Description

The pressures exerted by a number of gases are measured separately and when the gases are mixed together. They are shown to obey Dalton's law.

Apparatus

▼   One 250 cm³ Buchner flask with a one-hole rubber stopper.

▼   A **mercury** manometer (this can be made from about 1 m of flexible plastic tubing).

▼   One glass stopcock.

▼   One self-sealing rubber septum cap.

▼   One 20 cm³ (or larger) plastic syringe with hypodermic needle.

▼   Three or more 'gas bags' (see demonstration 18).

▼   A metre rule.

Chemicals

▼   Access to gases *eg* nitrogen, hydrogen, methane (natural gas) or others as available.

Method

Before the demonstration

Fill the gas bags with the different gases and label them. Set up the manometer and fill it with mercury. Arrange the manometer so that it can maintain a height difference of 15 cm between the mercury levels in the two arms without the mercury overflowing. It is recommended that the manometer is set up over a tray in case of any mercury spillage.

Insert one end of the glass stopcock in the one hole stopper and place the self-sealing rubber septum cap over the other end of the stopcock. If no such cap is available, a short length of rubber tubing sealed with a screw clip can be used instead. Connect the side arm of the Buchner flask to the manometer and place the stopper and stopcock assembly in the top of the Buchner flask. Clamp the Buchner flask to keep it stable.

NB: If the septum cap is brand new, it may be possible to omit the stopcock from the apparatus, but after a few uses the cap begins to leak slowly.

THE ROYAL
SOCIETY OF
CHEMISTRY

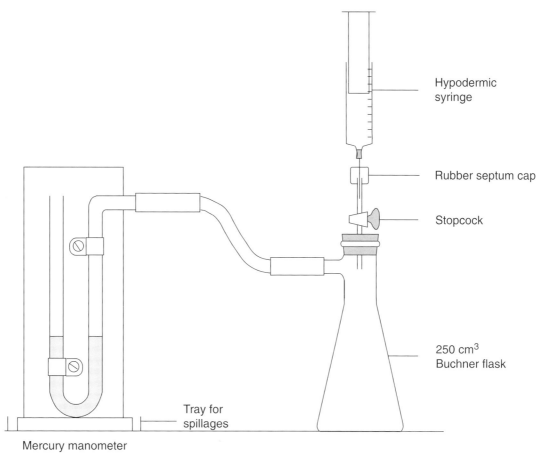

Hypodermic
syringe

Rubber septum cap

Stopcock

250 cm^3
Buchner flask

Tray for
spillages

Mercury manometer

Measuring gas pressures

The demonstration

Ensure that the mercury level is the same in both arms of the manometer by opening
the stopcock and removing the septum cap. Replace the septum cap. Insert the
syringe's hypodermic needle through the rubber cap of one gas bag and withdraw 20
cm^3 of gas. With the stopcock open, inject the gas into the Buchner flask. Close the
stopcock and record the difference in levels of the mercury between the two arms of
the manometer.

Re-equalise the mercury levels by opening the stopcock and repeat the procedure
with 15 cm^3 of the second gas. Then repeat the procedure with 10 cm^3 of the third
gas.

Now equalise the mercury levels and add the same quantities of each gas
successively, without equalising the mercury levels in between. Note the difference
in levels. This should be equal to the sum of the differences in level for the separately
added gases (within experimental error).

Visual tips

The mercury levels will be difficult for the audience to see. Get a member of the
audience to measure the height differences (to the nearest mm) using a ruler and
another student to record them on the blackboard.

THE ROYAL
SOCIETY OF
CHEMISTRY

Teaching tips

The fact that 20 cm³ of hydrogen exerts the same pressure as 20 cm³ of carbon dioxide may surprise many students and might lead to an interesting discussion.

Theory

Dalton's law of partial pressures states that a mixture of gases will exert a total pressure equal to the sum of the pressures exerted by each of the gases occupying the same container separately. In other words the pressure depends on the number, not the nature, of gas particles.

Extensions

Any number of gases may be used in any desired volumes, provided that the mercury does not overflow. Any gases may be used provided that they do not react together.

Safety

Wear eye protection.
    It is the responsibility of teachers doing this demonstration to carry out an appropriate risk assessment.

THE ROYAL
SOCIETY OF
CHEMISTRY

17. Anodising aluminium

Topic

Electrochemistry/electrolysis/industrial chemistry.

Timing

About 1 h.

Level

Upper secondary.

Description

A strip of aluminium is anodised and the thickened surface coating is dyed.

Apparatus

▼ Low voltage DC power pack – adjustable up to 15 V.

▼ 0–100 Ohm rheostat.

▼ 0–1 A ammeter.

▼ 0–15 V voltmeter.

▼ Connecting leads and crocodile clips.

▼ Retort stand with boss and clamp.

▼ Ruler (15 cm).

▼ One 1 dm^3 beaker.

▼ Four 250 cm^3 beakers.

▼ One 1 dm^3 conical flask.

Chemicals

The quantities given are for one demonstration.

▼ Aluminium foil approximately 50 cm x 50 cm.

▼ Dylon cold fabric dye (Camilla A 16). This is cherry red. (Some other colours may work as well.)

▼ 1 dm^3 of **sulphuric acid** (approximately 2 mol dm^{-3}).

▼ 250 cm^3 of **nitric acid** (approximately 1 mol dm^{-3}).

▼ 250 cm^3 of **sodium hydroxide** (approximately 1.5 mol dm^{-3}).

▼ 250 cm^3 of **propanone** (acetone).

▼ Strip of thin aluminium sheet approximately 12 cm x 3 cm.

▼ 2 cm^3 of **glacial ethanoic acid** (acetic acid).

Method

Before the demonstration

Line the inside of the sides of the 1 dm^3 beaker with a double thickness of aluminium foil. Fill the beaker with sulphuric acid. This should be at about 25 °C – adjust the temperature if necessary. Set up the electrical circuit shown in the figure. Make up the dye solution according to the instructions supplied (*ie* dissolve the contents of the tin in about 600 cm^3 of water) and add a few cm^3 of glacial ethanoic acid.

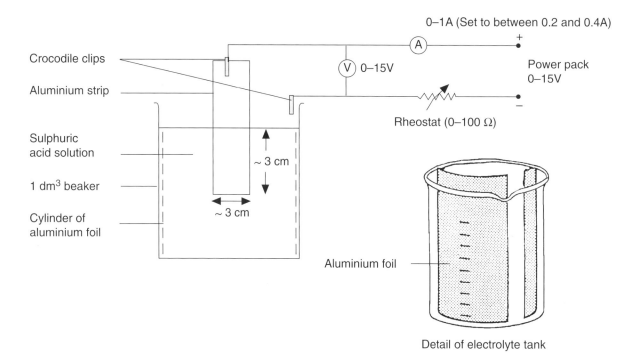

Apparatus for anodising aluminium

The demonstration

De-grease the aluminium strip by rubbing with a tissue soaked in propanone and then dipping the strip into a beaker of propanone for a few seconds and allowing to dry. From now on hold the aluminium by the top few cm only (where it will not be anodised).

Dip the bottom half of the aluminium strip into the sodium hydroxide solution in a beaker. Leave it until it begins to effervesce, indicating that the surface layer of oxide has been removed. (This will take about one minute.) Now remove the strip and dip the cleaned portion of it into the nitric acid for a few seconds to neutralise the alkali. Then rinse away the acid with water.

Clamp the strip so that the lower, cleaned, section is immersed in the sulphuric acid electrolyte and is in the centre of the cylinder of aluminium foil which forms the cathode. It must not touch the cathode.

Complete the circuit with crocodile clips making the aluminium strip positive and the foil negative. Now adjust the power pack and rheostat so that a current flows which gives a current density of 10–20 mA cm^{-2} of anode area immersed. For example if the anode has an area of 3 cm x 3 cm immersed, the area will be 3 x 3 x 2 cm^2 = 18 cm^2 (don't forget it has two sides!), so the current should be between 180 and 360 mA (0.18 and 0.36 A).

Leave to electrolyse for about 30 minutes, keeping an eye on the current and adjusting the rheostat if necessary to keep its value constant. (The current may tend to drop as the oxide layer thickens.)

When the electrolysis is complete, switch off the power and remove the aluminium strip. Rinse the strip in water. It will not look very different at this stage. Now dip the strip into about 200 cm^3 of the dye solution in a beaker. Make sure that

THE ROYAL
SOCIETY OF
CHEMISTRY

some of the non-anodised part of the strip is immersed as well as the treated section. Leave for about 15 minutes – longer immersion will produce a deeper colour. Some of those who trialled this demonstration left the strip in the dye overnight. Rinse to remove any dye which has not been absorbed. Dye will only be absorbed by the anodised section, which will turn a deep cherry red. If desired, seal the dye by immersing the dyed strip for a few minutes in water that is already boiling. This will make the colour less prone to rubbing off, but will wash out some of it. Many teachers may prefer to omit this procedure.

Visual tips

Large demonstration meters will be easy to see. Long connecting leads are useful to prevent the electrolysis tank becoming lost in a maze of wires.

Teaching tips

It would be wise to prepare something to fill in the half hour of electrolysis time and the 15 minutes dyeing time. The class could be asked to calculate the expected increase in mass of the anode or to discuss the chemical reactions involved. Have a selection of anodised objects such as saucepan lids available for the class to look at. The demonstration (No. 18) of the reactivity of aluminium without its normal oxide layer could be shown.

Some teachers may prefer to anodise some aluminium before the lesson to have some pieces ready to show the class.

Theory

Untreated aluminium has a layer of oxide about 10^{-8} m thick. This explains its apparent lack of reactivity. Anodising, invented in 1923, is used commercially to thicken this layer to 10^{-5} m to improve the metal's corrosion resistance. The relevant equations are:

cleaning:
$Al_2O_3(s) + 2OH^-(aq) + 3H_2O(l) \rightarrow 2Al(OH)_4^-(aq)$

once the oxide is removed:
$2Al(s) + 2OH^-(aq) + 6H_2O(l) \rightarrow 2Al(OH)_4^- + 3H_2(g)$

electrolysis at the anode:
$2Al(s) + 3H_2O(l) \rightarrow Al_2O_3(s) + 6H^+(aq) + 6e^-$

electrolysis at the cathode:
$6H^+(aq) + 6e^- \rightarrow 3H_2(g)$

electrolysis overall:
$2Al(s) + 3H_2O(l) \rightarrow Al_2O_3(s) + 3H_2(g)$

The oxide coating develops a positive charge by the reaction:

$Al_2O_3(s) + H_2O(l) \rightarrow Al_2O_3H^+(s) + OH^-(aq).$

Thus it attracts dyes that contain coloured anions. These are absorbed in the pores of the sponge-like oxide layer, where they can be trapped by heating the oxide to form an $Al_2O_3.H_2O$ seal.

THE ROYAL
SOCIETY OF
CHEMISTRY

Extensions

There are a great many variables in this experiment such as: electrolysis time, voltage, current density, concentration of electrolyte, temperature of electrolyte, temperature of dyebath and type of dye. Investigations of some of these could form interesting projects.

It is possible to measure the gain in mass of the anode by rinsing the aluminium strip with propanone and weighing it immediately before and immediately after electrolysis.

Safety

Wear eye protection.

Hydrogen is given off at the cathode during electrolysis, so avoid naked flames.

It is the responsibility of teachers doing this demonstration to carry out an appropriate risk assessment.

THE ROYAL
SOCIETY OF
CHEMISTRY

18. The real reactivity of aluminium

Topic

Metals/corrosion/reactivity series.

Timing

Less than 5 min.

Level

Pre-16.

Description

Sodium hydroxide and mercury(II) chloride solutions are used to remove the oxide layer from aluminium foil and the resulting vigorous reaction with air is shown.

Apparatus

▼   Two 250 cm³ beakers.

▼   Watch glass.

▼   Tongs or tweezers.

Chemicals

The quantities given are for one demonstration.

▼   2 g of **mercury(II) chloride** ($HgCl_2$).

▼   200 cm³ of 1 mol dm⁻³ **sodium hydroxide**.

▼   Aluminium cooking foil.

Method

Before the demonstration

Make a solution of 2 g of mercury(II) chloride in 200 cm³ of water.

The demonstration

Cut off about a 5 cm x 5 cm piece of aluminium cooking foil. Using tongs or tweezers, immerse the metal in the sodium hydroxide solution until it begins to effervesce (*ca* 1 minute). Holding it in the tweezers, rinse it with water and dip it in the mercury(II) chloride solution. Leave it for about 1 minute until the solution begins to turn slightly grey. Remove the foil using tweezers, rinse it, shake it dry and leave it on the watch glass. After about 1 minute, the foil will start to tarnish rapidly and will become coated with a pale grey layer of oxide. It will get hot, and steam will be produced as any remaining droplets of water evaporate.

Alternative method

Soak a little cotton wool or tissue in mercury(II) chloride solution and, using tweezers and plastic gloves, rub a little aluminium foil with it. The cleaned area soon tarnishes and becomes coated with a layer of white oxide.

THE ROYAL
SOCIETY OF
CHEMISTRY

Visual tips

Pass the remains of the foil round the class after it has reacted, along with an untreated piece for comparison. Place the treated piece in a closed glass petri dish because it may still be contaminated with the mercury(II) chloride solution.

Teaching tips

A milk bottle top can be used instead of foil.

Theory

Electrode potentials place aluminium between zinc and magnesium in reactivity, but the presence of a firmly-attached oxide layer means that it is usually unreactive enough to be used unpainted. Sodium hydroxide dissolves away the oxide coating on aluminium as sodium aluminate.

$$Al_2O_3(s) + 2NaOH(aq) + 3H_2O(l) \rightarrow 2NaAl(OH)_4(aq)$$

Once the oxide has gone, effervescence starts due to:

$$2Al(s) + 2NaOH(aq) + 6H_2O(l) \rightarrow 2NaAl(OH)_4(aq) + 3H_2(g)$$

Mercury(II) chloride forms an amalgam with the freshly exposed surface.

Safety

Wear eye protection and disposable plastic gloves.

Mercury(II) chloride is very toxic by inhalation, on contact with the skin and if swallowed. It may have cumulative effects. Mercury salts should not normally be disposed of down the sink. The solution could be kept in a closed container for further demonstrations. Alternatively, treat the mercury chloride with excess concentrated sodium hydroxide to precipitate yellow mercury hydroxide. Filter this off, place it in a sealed plastic bag and store it for professional disposal.

Do not throw treated aluminium foil in the rubbish bin. It may continue to react exothermically and ignite other rubbish. One method of disposal is to return it to the sodium hydroxide solution until it has dissolved.

It is the responsibility of teachers doing this demonstration to carry out an appropriate risk assessment.

THE ROYAL
SOCIETY OF
CHEMISTRY

19. Gas bags

Gas bags are a useful way of storing and dispensing small quantities of a variety of gases at atmospheric pressure for use in demonstrations (or class practicals). The gases are withdrawn from the bags using a syringe and hypodermic needle.

Construction

Method 1

You will need a plastic sandwich or freezer bag, a one hole rubber stopper (about 17 mm), 4 cm of glass tubing (to fit the stopper), a self sealing rubber septum cap or gas-syringe cap (to fit the glass tube), a cable tie (to fit around the stopper) and a 20 cm³ syringe with a hypodermic needle *(see Fig)*.

Insert the glass tube through the rubber stopper (CARE – use a cork borer) so that about 1 cm of tube protrudes from the top of the stopper. Gather the neck of the plastic bag around the stopper and secure it tightly with the cable tie. Check for leaks by closing the end of the glass tube with a rubber cap, immersing the assembly in a bowl of water and squeezing. Squeeze the air from the bag. It can now be filled with gas from any suitable source such as a cylinder or chemical generator. Fill the bag and squeeze out the gas two or three times to ensure that any air is flushed out. Once filled, close the glass tube with the rubber cap. Gas can now be drawn into the syringe by injecting the hypodermic needle through the cap and sucking gas into the syringe.

An even simpler alternative is to use a party balloon instead of the plastic bag. However these can only be inflated with gas from a source at relatively high pressure such as a cylinder.

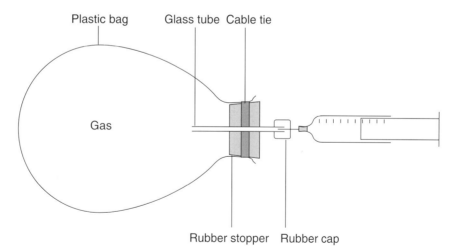

Apparatus for method 1

Method 2

You will need an empty wine box, a one hole rubber stopper (to fit the mouth of the wine box once the tap has been removed), 4 cm of glass tubing (to fit the stopper), a self sealing rubber septum cap or gas-syringe cap (to fit the glass tube) and a 20 cm³ syringe with a hypodermic needle.

Using a sharp knife remove the cardboard top of the wine box (which contains the handle). Take care not to damage the bag inside. Using a junior hacksaw, saw off

THE ROYAL
SOCIETY OF
CHEMISTRY

the tap of the wine box. This is usually made of black plastic and fits into a grey plastic sleeve. Rinse out the bag of the wine box and allow to dry. Insert the glass tube through the rubber stopper (CARE – use a cork borer) so that about 1 cm of tube protrudes from the top of the stopper. Fit the stopper into the hole created by removing the tap and squeeze the air out of the bag. The bag can now be filled with gas from any suitable source such as a cylinder or chemical generator. Fill the bag and squeeze out the gas two or three times to ensure that any air is flushed out. Once filled, close the glass tube with the rubber cap. Gas can now be withdrawn into the syringe by injecting the hypodermic needle through the cap and sucking gas into the syringe.

NB: Cable ties are plastic ties with a non-releasable ratchet action used for clipping runs of cable together (see below). They can be used with or without a gun. They are obtainable from DIY or motor accessory stores.

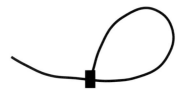

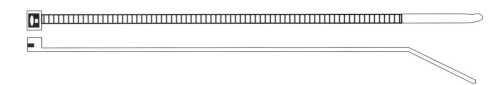

THE ROYAL
SOCIETY OF
CHEMISTRY

20. The hydrogen peroxide/potassium iodide clock reaction

Topic

Reaction rates/kinetics or general interest/problem-solving.

Timing

1 min to 1 h, depending on what is attempted.

Level

Pre-16, as an introduction to reaction rates. Post-16, for reaction rates/mechanism. Any, for general interest or as a problem-solving activity.

Description

A solution of hydrogen peroxide is mixed with one containing potassium iodide, starch and sodium thiosulphate. After a few seconds the colourless mixture suddenly turns dark blue.

Apparatus

▼   One 1 dm³ beaker.

▼   Two 400 cm³ beakers.

▼   One 1 dm³ volumetric flask.

▼   Magnetic stirrer and follower (optional).

▼   Stopclock.

Chemicals

The quantities given are for one demonstration.

▼   0.2 g of soluble starch.

▼   30 cm³ of **glacial ethanoic acid** (glacial acetic acid, CH_3CO_2H).

▼   4.1 g of anhydrous sodium ethanoate (sodium acetate, CH_3CO_2Na).

▼   50 g of potassium iodide (KI).

▼   9.4 g of sodium thiosulphate-5-water ($Na_2S_2O_3.5H_2O$).

▼   500 cm³ of 20 volume hydrogen peroxide solution ($H_2O_2(aq)$).

▼   2 dm³ of deionised water.

Method

Before the demonstration

Make up solution A as follows:

Make a paste of 0.2 g of soluble starch with a few drops of water in a beaker. Pour onto this 100 cm³ of boiling water and stir. Pour the resulting solution into a 1 dm³ beaker and dilute to 800 cm³. Add 30 cm³ of glacial ethanoic acid, 4.1 g of sodium ethanoate, 50 g of potassium iodide and 9.4 g of sodium thiosulphate. Stir until all the solids have dissolved, and, when it has cooled to room temperature, pour the mixture into a 1 dm³ volumetric flask and make up to 1 dm³ with water.

Make up solution B as follows:
Dilute 500 cm³ of 20 volume hydrogen peroxide to 1 dm³ with water.
Both solutions are colourless although A will be slightly cloudy.

The demonstration

Measure 100 cm³ of solution A and 100 cm³ of solution B in separate beakers. Pour one into the other and swirl vigorously to mix. After about 20 seconds at room temperature the mixture will suddenly turn dark blue.

The appearance of the blue colour may be timed and the effect of varying the temperature of both solutions before mixing may be shown. (A temperature rise/fall of 10 °C roughly doubles/halves the rate as expected.)

Halving the concentration of solution B (50 cm³ of B + 50 cm³ of water) will double the time for the blue colour to appear as the reaction rate is halved. However, halving the concentration of solution A (50 cm³ of A + 50 cm³ of water) will cause no change to the time period because the rate has been halved and the amount of thiosulphate has been halved too. Hence you only need to make half the amount of iodine to mop up the thiosulphate and this is made half as fast: the two effects cancel each other out. This will probably surprise some students and may need to be explained carefully.

Visual tips

A white background will help so that the impact of the sudden and spectacular colour change is not lost. Scaling up the volumes of solution that are mixed may help in a large room.

Because of the time required for mixing, it is not easy to know when to start the clock to time the reaction accurately. One way round this is to stir solution A on a magnetic stirrer and add B to it, starting the clock when B is half poured. An assistant to do the timing may help. There is no warning of when the blue colour is about to appear.

Teaching tips

It may help understanding if the students are already familiar with the reactions of starch and iodine and iodine and sodium thiosulphate. It may be worth demonstrating these beforehand.

The time for the blue colour to appear can be adjusted by varying the amount of thiosulphate in solution A so a 'clock' of any desired time interval can be produced. If the demonstration is being done for entertainment, the imaginative teacher will be able to think up some suitable patter.

Theory

The basic reaction is:

$$H_2O_2(aq) + 2I^-(aq) + 2H^+(aq) \rightarrow I_2(aq) + 2H_2O(l).$$

This reaction is the rate determining step and is first order with respect to both H_2O_2 and I^-.

As soon as the iodine is formed, it reacts with the thiosulphate to form tetrathionate ions and recycles the iodide ions by the fast reaction:

$$2S_2O_3^{2-}(aq) + I_2(aq) \rightarrow S_4O_6^{2-}(aq) + 2I^-(aq)$$

As soon as all the thiosulphate is used up, free iodine (or, strictly, I_3^- ions) remains and reacts with the starch to form the familiar blue-black complex.

THE ROYAL
SOCIETY OF
CHEMISTRY

Notes

1. Hydrogen peroxide is capable of oxidising thiosulphate ions to tetrathionate ions but the reaction is too slow to affect this demonstration.

2. The ethanoic acid/sodium ethanoate is added to buffer the pH.

Extensions

The demonstration could lead to an investigation of the factors affecting reaction rates. This could be planned conventionally by, say, a group of post-16 students who have covered the theory of the reaction or as a problem-solving exercise by a group who do not know the details but simply ask questions such as 'What happens if we dilute A and/or B?' The unexpected result of diluting A will add spice to this.

Some teachers may wish to use a method in which the sodium thiosulphate is omitted from solution A and added separately in a third solution. This means that the unusual result on diluting solution A is avoided.

Safety

Wear eye protection.

The solution remaining after the experiment can be flushed down the sink with plenty of water.

It is the responsibility of teachers doing this demonstration to carry out an appropriate risk assessment.

THE ROYAL
SOCIETY OF
CHEMISTRY

21. Phenol/methanal polymerisation

Topic

Polymers (condensation).

Timing

About 5 min.

Level

Introductory polymerisation or post-16 to introduce discussion of reaction mechanism.

Description

A pink thermosetting polymer is produced by adding sulphuric acid to an acidified solution of phenol in formalin.

Apparatus

▼ One 250 cm^3 beaker.

▼ One 100 cm^3 measuring cylinder.

▼ One screw-top coffee jar.

▼ Stirring rod.

▼ Access to a fume cupboard.

Chemicals

The quantities given are for one demonstration.

▼ 20 g of **phenol** (C_6H_5OH).

▼ 25 cm^3 of **formalin** (a 37 % solution of methanal (formaldehyde, HCHO) in water).

▼ 55 cm^3 of **glacial ethanoic acid.**

▼ 30 cm^3 of **concentrated sulphuric acid**.

Method

Before the demonstration

Add 30 cm^3 of concentrated sulphuric acid to 30 cm^3 of water slowly and with stirring. Cool the resulting diluted acid to room temperature.
Weigh out 20 g of phenol.

The demonstration

Work in a fume cupboard and wear disposable plastic gloves. Pour 25 cm^3 of formalin into a coffee jar and add 55 cm^3 of glacial ethanoic acid. Then stir in 20 g of phenol until it has dissolved. Now add the 60 cm^3 of diluted sulphuric acid stirring continuously. The mixture will turn pale yellow and after a few seconds will suddenly go an opaque pink colour. Over the next minute or so a pink solid will gather around the stirring rod. A lot of heat is evolved.

Pour off the remaining milky liquid and rinse the pink polymer several times with water. Show that this polymer is hard by poking it with a spatula. Break off a small piece and heat it in a Bunsen flame, holding it with tongs. The material will char but

THE ROYAL
SOCIETY OF
CHEMISTRY

not soften, showing that the polymer is thermosetting.

If you intend to pass the material round the class, the screw top should be placed on the jar as the product will be contaminated by unreacted starting materials. It may be necessary to break off the stirring rod.

Visual tips

When demonstrating in a fume cupboard, take care that your body does not obstruct the students' view. A mobile fume cupboard would be ideal if available.

Teaching tips

The polymer produced is Bakelite, the first genuine synthetic polymer (as opposed to modified natural polymers such as Celluloid, modified cellulose). It was developed by Leo Bakeland in the early 1900s, and it is used to make articles such as electrical fittings.

Theory

The basic reaction is a condensation polymerisation in which water is eliminated.

$$OH \quad + \quad H-\overset{\overset{O}{\|}}{C}-H \quad + \quad OH \quad \longrightarrow \quad OH-CH_2-OH \quad + H_2O \; etc$$

The product has considerable crosslinks, including some $-CH_2OCH_2$ linkages.

Further details

Some teachers may prefer to scale down the quantities to half those described above.

Try moulding an object with a simple shape by pouring the yellowish liquid obtained after adding the sulphuric acid into a suitable mould.

Safety

Wear eye protection and plastic gloves and work in a fume cupboard.

If the polymer is made in a throwaway screw-top container, this will aid disposal as solid waste.

It is the responsibility of teachers doing this demonstration to carry out an appropriate risk assessment.

THE ROYAL
SOCIETY OF
CHEMISTRY

22. The 'blue bottle' experiment

Topic

General interest, problem-solving, chemical reactions.

Timing

1 min upwards.

Level

Primary or lower secondary.

Description

A colourless solution in a flask is shaken. It turns blue and then gradually back to colourless. The cycle can be repeated many times.

Apparatus

▼   One 1 dm³ conical flask with stopper.

▼   Access to a nitrogen cylinder (optional).

▼   Access to a fume cupboard (optional).

Chemicals

The quantities given are for one demonstration.

▼   8 g of **potassium hydroxide** or 6 g of **sodium hydroxide**.

▼   10 g of glucose (dextrose).

▼   0.05 g of methylene blue.

▼   50 cm³ of **ethanol**.

Method

Before the demonstration

Make a solution of 0.05 g of methylene blue in 50 cm³ of ethanol (0.1 %).

Weigh 8 g of potassium hydroxide or 6 g of sodium hydroxide into a 1 dm³ conical flask. Add 300 cm³ of water and 10 g of glucose and swirl until the solids are dissolved. Add 5 cm³ of the methylene blue solution. None of the quantities is critical. The resulting blue solution will turn colourless after about one minute. Stopper the flask.

The demonstration

Shake the flask vigorously so that air dissolves in the solution. The colour will change to blue. This will fade back to colourless over about 30 seconds. The more shaking, the longer the blue colour will take to fade. The process can be repeated for over 20 cycles. After some hours, the solution will turn yellow and the colour changes will fail to occur.

Visual tips

A white laboratory coat provides the ideal background.

Teaching tips

On a cold day, it may be necessary to warm the solution to 25–30 °C or the colour changes will be very slow.

THE ROYAL
SOCIETY OF
CHEMISTRY

The demonstration can be used to start a discussion on what is causing the colour changes. Students' suggestions can be tried out as far as is practicable.

Theory

Glucose is a reducing agent and in alkaline solution will reduce methylene blue to a colourless form. Shaking the solution admits oxygen which will re-oxidise the methylene blue back to the blue form.

Extensions

To confirm that oxygen is responsible for the colour change, nitrogen can be bubbled through the solution for a couple of minutes to displace air from the solution and the flask. If the stopper is now replaced and the bottle shaken, no colour change will occur. Reintroducing the air by pouring the solution into another flask and shaking will restore the system.

Natural gas can be used (in a fume cupboard) if nitrogen is not available.

Some teachers may wish to present this experiment as a magic trick. The colour change can be brought about by simply pouring the solution from a sufficient height into a large beaker.

This experiment can be a popular open-day activity. If visitors are to be allowed to shake the bottle themselves it might be wise to use a plastic screw-top pop bottle to eliminate the risk of the stopper coming off or the bottle being dropped and broken. The solution does not appear to interact with the plastic over a period of a day but it would be wise to try out the bottle you intend to use.

Redox indicators other than methylene blue can be used. In each case add the stated amount of indicator to the basic recipe of 10 g of glucose and 8 g of potassium hydroxide in 300 cm^3 of water.

1. **Phenosafranine.** This is red when oxidised and colourless when reduced. Use about 6 drops of a 0.2 % solution in water for a bottle that goes pink on shaking and colourless on standing. The initial pink colour takes some time to turn colourless at first. A mixture of phenosafranine (6 drops) and methylene blue (about 20 drops of the 0.1 % solution in ethanol) gives a bottle which will turn pink on gentle shaking through purple with more shaking and eventually blue. It will reverse the sequence on standing.

2. **Indigo carmine.** Use 4 cm^3 of a 1 % solution in water. The mixture will turn from yellow to red-brown with gentle shaking and to pale green with more vigorous shaking. The changes reverse on standing.

3. **Resazurin.** Use about 4 drops of a 1 % solution in water. This goes from pale blue to a purple-pink colour on shaking and reverses on standing. On first adding the dye, the solution is dark blue. This fades after about one minute.

Mixtures of the above dyes can also be used.

Further details

A. G. Cook, R. M. Tolliver and J. E. Williams, *J. Chem. Ed.,* 1994, **71**, 160. The article *The blue bottle experiment revisited* gives some details of the reaction mechanism and alternative dyes.

Safety

Wear eye protection.

It is the responsibility of teachers doing this demonstration to carry out an appropriate risk assessment.

THE ROYAL
SOCIETY OF
CHEMISTRY

23. The 'Old Nassau' clock reaction

Topic

Reaction rates.

Timing

Less than 5 min.

Level

Post-16 if the reactions are to be discussed. Any for interest/entertainment.

Description

Three colourless solutions containing iodate(V) ions, hydrogensulphite ions, mercury(II) ions and starch are mixed. After a few seconds the solution suddenly turns orange with a precipitate of mercury(II) iodide and a few seconds later suddenly turns black with the starch-iodine complex. Orange and black are the colours of the House of Nassau.

Apparatus

▼   Three 1 dm³ graduated flasks.

▼   Three 250 cm³ beakers or flasks.

▼   Stopwatch or stopclock (optional).

Chemicals

The quantities given are for one demonstration.

▼   4 g of soluble starch.

▼   13. 7 g of **sodium metabisulphite** ($Na_2S_2O_5$).

▼   3 g of **mercury(II) chloride** ($HgCl_2$).

▼   15 g of **potassium iodate(V)** (potassium iodate, KIO_3)

▼   About 3 dm³ of deionised water.

Method

Before the demonstration
Make up three solutions in the graduated flasks as follows.

A.   Make a paste of 4 g of soluble starch with a few drops of water. Pour onto this 500 cm³ of boiling water and stir. Cool to room temperature, add 13.7 g of sodium metabisulphite and make up to 1 dm³ with water.

B.   Dissolve 3 g of mercury(II) chloride in water and make the solution up to 1 dm³ with water.

C.   Dissolve 15 g of potassium iodate(V) in water and make the solution up to 1 dm³ with water.

The demonstration
Mix 50 cm³ of solution A with 50 cm³ of solution B. Pour this mixture into a beaker containing 50 cm³ of solution C. After about 5 seconds the mixture will turn an

THE ROYAL
SOCIETY OF
CHEMISTRY

opaque orange colour as insoluble mercury(II) iodide precipitates. After a further 5 seconds, the mixture suddenly turns blue-black as a starch-iodine complex is formed.

Visual tips

A white background helps visibility. Scale up the quantities if the reaction is being shown in a large room.

Theory

Sodium metabisulphite reacts with water to form sodium hydrogensulphite:

$$Na_2S_2O_5(aq) + H_2O(l) \rightarrow 2NaHSO_3(aq).$$

Hydrogensulphite ions reduce iodate(V) ions to iodide ions:

$$IO_3^-(aq) + 3HSO_3^-(aq) \rightarrow I^-(aq) + 3SO_4^{2-}(aq) + 3H^+(aq)$$

Once the concentration of iodide ions is large enough that the solubility product of HgI_2 (4.5×10^{-29} mol^3 dm^{-9}) is exceeded, orange mercury(II) iodide solid is precipitated until all of the Hg^{2+} ions are used up (provided that there is an excess of I^- ions).
      If there are still I^- and IO_3^- ions in the mixture, the reaction

$$IO_3^-(aq) + 5I^-(aq) + 6H^+(aq) \rightarrow 3I_2(aq) + 3H_2O(l)$$

takes place and the blue starch-iodine complex is formed.

Extensions

Diluting all of the solutions by a factor of two increases the time taken for the colour changes to occur. Using a smaller volume of solution B speeds up the reaction. The effects of changing the amounts and concentrations of the various reactants cannot always be predicted simply because of the complexity of the system. For example if the volume of solution B is doubled, the appearance of the orange colour is delayed and the blue colour fails to appear at all.

Further details

A full account of the reaction can be found in B. Z. Shakhashiri, *Chemical demonstrations: A handbook for teachers of chemistry*, Volume 4. Wisconsin, US: The University of Wisconsin Press, 1992.

Safety

Wear eye protection and plastic gloves.
      To dispose of reaction mixtures after the demonstration, filter off the insoluble mercury (II) iodide and place the filter paper in a sealed plastic bag. Any unused solution B should be treated with excess sodium hydroxide to precipitate insoluble orange mercury hydroxide. This should be filtered off and the filter paper placed in a sealed plastic bag. The mercury residues should then be retained for professional disposal.
      It is the responsibility of teachers doing this demonstration to carry out an appropriate risk assessment.

THE ROYAL
SOCIETY OF
CHEMISTRY

24. Gas chromatography

Topic

Analytical techniques/separation of mixtures.

Timing

10 min or more.

Level

Post-16 or possibly pre-16.

Description

Volatile hydrocarbons are injected into a column containing washing powder.
Natural gas is used as a carrier gas and burnt at a jet. The elution of the samples is
detected by changes in the size and luminosity of this flame, which can be compared
with a reference flame.

Apparatus

▼   One U-tube approximately 12 cm in length, with side arms .

▼   One rubber stopper to fit the U-tube mouth.

▼   One rubber septum cap to fit the U-tube mouth.

▼   Three 1 cm^3 plastic syringes with hypodermic needles.

▼   One 1 dm^3 beaker.

▼   Plastic and/or rubber tubing – see diagram.

▼   Glass funnel (optional).

▼   Stopclock.

Chemicals

▼   A few cm^3 of **pentane**.

▼   A few cm^3 of **hexane**.

▼   A few grams of Surf® washing powder.

Method

Before the demonstration

Dry the washing powder overnight in an oven. Fill the U-tube with washing powder
to just below the level of the side arms. Stopper one mouth of the U-tube and fit a
self-sealing rubber septum cap over the other. Connect one side arm to the gas tap
with rubber tubing. The other side arm is connected with plastic or rubber tubing to a
1 cm^3 plastic syringe which has had the handle end removed using scissors or a sharp
knife. Attach a hypodermic needle to the syringe and clamp this vertically to form a
jet at which the gas can be burnt. Immerse the U-tube in a beaker full of water that
has just boiled (see Fig). Make a second jet with a cut-off syringe and hypodermic
needle. Attach this by rubber tubing to a second gas tap. This will be used to provide
a reference flame. Clamp this jet at the same height as, and close to, the first one.

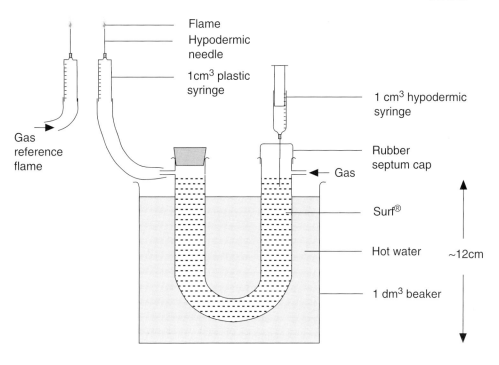

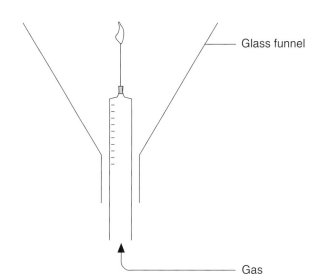

Separation of gas mixtures

The demonstration

Turn on the gas to the column and the reference jets. Light the jets and adjust them so that they have identical, almost non-luminous, flames about 2 mm high – screw clips on the gas tubes can help to achieve fine control. (It will take a short time for the air to be displaced from the column before the jet can be lit.) It may be necessary to improvise a draught shield to stop these small flames from going out. Take up about 0.02 cm³ of pentane in a hypodermic syringe. Insert the needle through the septum cap and into the Surf® and inject the pentane onto the column. Start the stopclock. Observe the flame carefully and note when it becomes significantly taller and more

THE ROYAL
SOCIETY OF
CHEMISTRY

luminous than the reference flame. This will take roughly 40 seconds. Note also when the flame returns to normal (about 50 seconds later). Repeat this process with hexane instead of pentane. This will come through after about 80 seconds and affect the flame for about a further 70 seconds.

Visual tips

A black background seems to be best for observing the flames.

Teaching tips

After establishing the times for each compound to come through the column, try an 'unknown' – either hexane or pentane – and try to identify which it is. It is difficult to resolve a mixture reliably because the 'tail' of the pentane flame tends to mingle with the hexane one. Point out that real columns and detection systems can effect much better separations.

Theory

An equilibrium is set up between pentane adsorbed on the column material and that in the gas phase. Hexane is more strongly adsorbed on the column and is therefore carried more slowly by the carrier gas.

Extensions

Try heptane. Try other compounds of similar volatility to pentane and hexane. Investigate the effect of the following variables: temperature, column length, column cross-sectional area, column filling material (kieselguhr works well, other washing powders could be used), volume of compound injected, gas flow rate *etc.*

Further details

One teacher involved in the trials reported that the following filling materials were all effective: Surf Micro®, Radion®, Ariel Ultra®, Persil®, acid washed sand (40–100 mesh), table salt.

Retention times depend on the details of the apparatus and may differ from those reported here.

Safety

Wear eye protection.

It is the responsibility of teachers doing this demonstration to carry out an appropriate risk assessment.

THE ROYAL
SOCIETY OF
CHEMISTRY

25. Bubbles that float and sink

Topic

This demonstration will probably be done for entertainment but it touches on a number of areas of chemistry – density of gases, diffusion, solubility of gases, sublimation and combustion.

Timing

About 10 min.

Level

Almost any.

Description

Dry ice is used as a source of carbon dioxide gas with which to blow bubbles which sink in air. Hydrogen (or methane) is used to blow bubbles which float (and which can be ignited). With patience, bubbles can be blown which first sink and then float.

Apparatus

▼ Expanded polystyrene container such as a cool box or the containers in which Winchester bottles are sometimes supplied.

▼ One 1 dm³ conical flask with a two-holed rubber stopper to fit.

▼ Glass and plastic tubing – *(Fig. 1)*.

▼ Small (approximately 2 cm diameter) plastic funnel.

▼ One 100 cm³ beaker.

▼ Hydrogen cylinder and regulator. (Methane from the gas tap can be used instead.)

▼ Insulating gloves with which to handle the dry ice.

▼ A candle taped to a stick about 1 m long (a metre rule would do).

Chemicals

The quantities given are for one demonstration.

▼ About 100 g of dry ice (solid carbon dioxide).

▼ A few cm³ of Teepol or washing up liquid, or special bubble mix – see below.

▼ A few cm³ of glycerol (glycerine).

Method

Before the demonstration

Make a bubble pipe by gluing a length of absorbent material, such as loosely woven shoelace, around the inner rim of a plastic funnel *(Fig. 1)*. This acts as a wick that soaks up and retains bubble mixture so that several bubbles can be blown without re-dipping the pipe into the bubble mixture.

Make about 50 cm³ of bubble mixture by mixing roughly 5 cm³ of Teepol, 5 cm³ of glycerol and 40 cm³ of water. If another detergent is used, it is worth doing some preliminary experimentation to find a mixture that produces robust bubbles with the gases used. Alternatively a bubble mixture can be obtained from the Bristol

Exploratory, Bristol Old Station, Temple Meads, Bristol, BS1 6QU, Tel: 01272 252008. The cost (1994) is 95 p + P&P. This is based on an US detergent, Dawn. Bubble mixture bought from toyshops does not seem to be very effective for this demonstration.

Dry ice can be obtained from a local university, higher education institution or industry. It can be stored and transported in an expanded polystyrene box and can be kept for a few hours.

Wind some insulating tape around the flask as a precaution against explosion. Fit the conical flask with the stopper and tubes as shown in the diagram and connect it to the hydrogen cylinder or gas tap. Half fill the flask with water that has been warmed to a few degrees above room temperature.

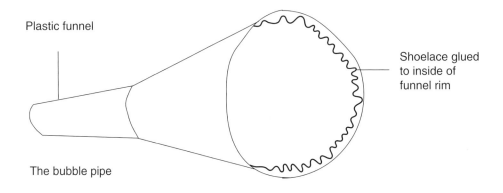

Fig. 1 The bubble pipe

The demonstration

Dip the bubble pipe into the bubble mixture so that the wick is soaked. Using gloves put a few small pieces of dry ice into the flask and re-stopper it. A 'fog' will form in the flask and fog-filled bubbles of carbon dioxide will form. These will sink rapidly to the ground. A plastic tray on the floor will reduce mess and should allow some bubbles to remain on it without bursting. These will shrink as the water-soluble carbon dioxide diffuses out of the bubble faster than the less soluble air diffuses in.

When the dry ice is used up, turn on the hydrogen cylinder and blow hydrogen bubbles (*Fig. 2*). These will float up to the ceiling and can be ignited (noisily) with the candle on the stick. Take great care not to ignite bubbles while they are still on the pipe to avoid the risk of igniting the flask of hydrogen.

Now add more dry ice to the flask and blow bubbles with a mixture of hydrogen and carbon dioxide. These will sink or float depending on the proportions of the two gases. With practice and some patience it is possible to blow bubbles that start to sink and then, as the carbon dioxide diffuses out, float upwards. The best method is to start by blowing pure CO_2 bubbles, gradually increasing the proportion of hydrogen until the bubble just starts to pull downwards while still attached to the pipe. The bubble can then be released by shaking it gently from the pipe.

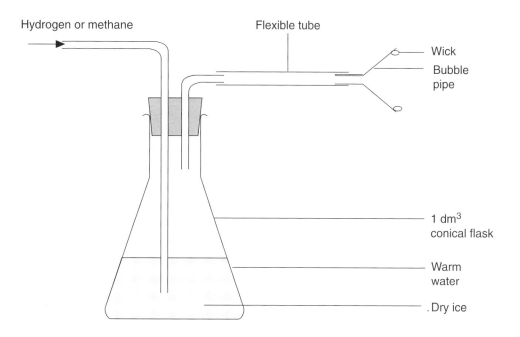

Fig. 2 Blowing gas bubbles

Visual tips

Clamp the bubble blowing apparatus on the edge of the bench to allow the bubbles to fall to the floor.

Teaching tips

You will probably need a responsible assistant if you wish to ignite the bubbles. Students will probably enjoy this, but they must understand the risk of igniting bubbles still attached to the pipe.

Students will probably know of the use of dry ice to generate fog at rock concerts. The fog is not carbon dioxide, it is droplets of water condensed from the air by the cold dry ice (approximately –80 °C).

If the audience is appropriate, point out that the shrinkage of CO_2 bubbles is the opposite of what might be expected at first sight. If it was not for the solubility factor, lighter air molecules might be expected to diffuse into the bubble faster than CO_2 escapes.

The sublimation of CO_2 is worth pointing out.

Safety

Wear eye protection.

Tape the flask in case of explosion.

Take care not to ignite hydrogen bubbles while still on the bubble pipe.

Wear gloves (insulating ones, not rubber) to handle dry ice – it can cause frostbite burns.

Take care that tubing does not become pinched – plastic tubing is more rigid than rubber and is less prone to this.

It is the responsibility of teachers doing this demonstration to carry out an appropriate risk assessment.

Acknowledgement

This demonstration was adapted from an idea developed at the Bristol Exploratory.

THE ROYAL
SOCIETY OF
CHEMISTRY

26. Liquid nitrogen demonstrations

Topic

These demonstrations cover a variety of topics, but they are described together because institutions will usually have occasional access to liquid nitrogen and will probably wish to do all of the demonstrations at once. They could be done as part of a lecture demonstration for interest and stimulation. The topics covered include gas laws, liquefaction of gases, fractional distillation, combustion and bonding (the paramagnetism of oxygen).

Timing

About 1 h for all of the demonstrations.

Level

Almost any group of students will find some relevant chemistry.

Description

1.   Various materials are immersed in liquid nitrogen to demonstrate the effects of low temperature on their physical properties.

2.   Liquid air is made and distilled to show that it contains oxygen and nitrogen.

3.   Liquid oxygen is made and its colour, magnetism and ability to support combustion are shown.

4.   Solid carbon dioxide is made.

5.   Liquid nitrogen is boiled and the gas used to inflate a balloon to show the volume increase.

Apparatus

▼   One vacuum flask – about 1.5 dm³ is suitable. Ordinary household flasks appear to be suitable. It is helpful if more than one flask is available.

▼   Expanded polystyrene box such as those in which Winchester bottles are sometimes supplied.

▼   About 60 cm of copper tube as used for small-bore central heating systems. This can be bent by hand.

▼   Powerful permanent magnet such as an Eclipse Major.

▼   2–3 m of cotton thread or string.

▼   Two party balloons.

▼   Wooden spills.

▼   Rubber tubing.

▼   Two pyrex test-tubes 12 x 150 mm.

▼   Plastic fish tank (optional).

Chemicals

▼   About 1.5 dm³ of liquid nitrogen will be enough to do all of the demonstrations described.

THE ROYAL
SOCIETY OF
CHEMISTRY

▼   Lead sheeting, about 1–2 mm thick.

▼   A few cm³ of **mercury**.

▼   Access to an **oxygen** cylinder and regulator.

▼   Access to a **hydrogen** cylinder and regulator.

▼   Access to a carbon dioxide cylinder with regulator.

▼   A little washing up liquid and some food dye (optional).

▼   Bubble mixture from a toyshop (or a mixture of 5 cm³ of Teepol, 5 cm³ of glycerol and 40 cm³ of water).

▼   One banana and/or flower.

Method

The demonstration

1.   Pour some liquid nitrogen into an expanded polystyrene box and immerse a variety of items into it.

   a)   Mercury (in a small test-tube) solidifies.

   b)   Rubber or plastic tubing goes brittle and can be snapped in gloved hands or shattered with a sharp blow.

   c)   A strip of lead becomes rigid and will 'ring' when tapped with another metal.

   d)   A banana, for example, will become brittle and will shatter if it is hit on the bench. Take care to clear up all the shattered pieces – they become very soggy on melting!

   e)   Flowers or leaves will become brittle.

   f)   A balloon blown up with air (by mouth) will shrink as the air liquefies. Liquid air can be seen and felt inside the balloon. The rubber attains a texture rather like a crisp packet. The change is fully reversible. A balloon blown up with hydrogen from a cylinder shrinks to about 1/3 of its original size because in this case the gas will not liquefy. The balloon expands back to its original size when it is removed. If a suitably sized balloon is used, the balloon will sink to the floor when it is cold and shrunken and will float as it warms up and expands. These are dramatic examples of the gas laws and deviations from them.

   g)   A soap bubble can be held on a wire 'wand' over a container of liquid nitrogen. It will freeze.

2.   Place a clean dry test-tube in liquid nitrogen so that the lower half is immersed. Leave this for about 15 minutes until about 2 cm³ of liquid air has condensed in the tube. Take care that the mouth of the test-tube protrudes from the container as shown in *Fig 1a*. If the arrangement in *Fig 1b* is used, liquid nitrogen will be obtained rather than liquid air because the space above the liquid nitrogen will be filled with nitrogen gas, not air. Remove the test-tube from the liquid nitrogen and stand it in a test-tube rack. As it warms up, nitrogen (T_b = 77 K) will boil off first and will extinguish a lighted or glowing spill. After a few seconds, oxygen (T_b = 90 K) will start to boil off and

THE ROYAL
SOCIETY OF
CHEMISTRY

will relight a glowing spill. This illustrates the industrial fractional distillation of liquid air.

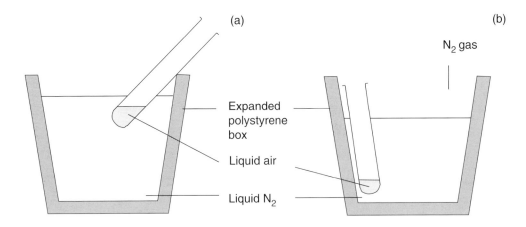

Fig.1 (a) Right and (b) wrong ways of test-tube immersion

3. Make a U-tube from copper pipe to fit the vacuum flask (*Fig. 2*). Place this in the liquid nitrogen and pass a stream of oxygen through it from a cylinder. After a couple of minutes, **liquid oxygen** will have condensed inside the U-tube and can be poured into a test-tube which has been previously cooled in liquid nitrogen. Note the pale blue colour. Hang the test-tube containing the liquid oxygen by cotton thread from the ceiling of the laboratory and bring a powerful magnet up to it. The tube will be pulled away from the vertical slightly showing that oxygen is paramagnetic. This does not occur with liquid nitrogen.

    Any remaining liquid oxygen can be poured onto a tissue on a heat-proof mat or sand tray. The tissue will burn vigorously when ignited with a burning spill.

4. a) Inflate a balloon with carbon dioxide gas from a cylinder. Attach the balloon securely over the mouth of a test-tube and immerse the tube in liquid nitrogen. The balloon will deflate as the gas solidifies and the tube will be filled partially with solid carbon dioxide.

    b) Inflate a balloon with carbon dioxide gas from a cylinder. Place the balloon in a container of liquid nitrogen, or pour liquid nitrogen over it. The balloon will collapse as the CO_2 solidifies. Cut the balloon open with scissors to reveal a frost of dry ice inside.

5. Pour 1 cm³ of water into a test-tube and mark the level that it reaches. Empty and dry the tube and cool it by immersion in liquid nitrogen. Pour about 2 cm³ of liquid nitrogen into the tube and hold the tube in liquid nitrogen to prevent the liquid in the tube from boiling away until ready for the next step. Take the test-tube out of the liquid nitrogen and allow the liquid inside to boil away until the level reaches the 1 cm³ mark. Now secure a balloon over the mouth of the tube and let the remaining 1 cm³ of liquid nitrogen boil away, inflating the balloon. When all the liquid has boiled away, remove the balloon and tie it, taking care not to lose any gas. Estimate the volume of the inflated balloon from its diameter or by immersing it in a bowl of water and measuring the increase in water level. One cm³ of liquid nitrogen gives about 2 dm³ of gas.

THE ROYAL
SOCIETY OF
CHEMISTRY

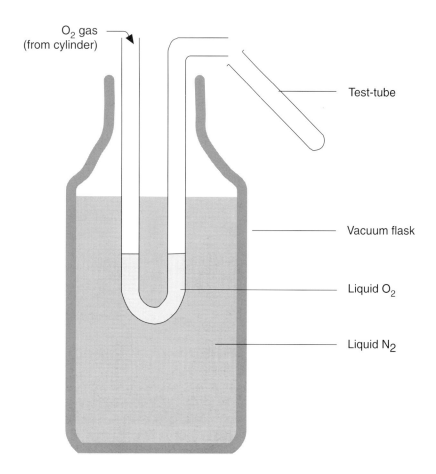

Fig. 2 Liquefying oxygen

Any remaining liquid nitrogen can be disposed of as follows.

a)   Pour some into a plastic fish tank of water containing a little washing up liquid and, if desired, a few drops of food dye. A spectacular fog will be produced and frozen bubbles will be left behind.

b)   Pour a little on the floor close to the feet of the audience to allow them to experience the coldness as it boils away.

Teaching tips

Point out that the liquid nitrogen is constantly boiling at room temperature and pressure as it is almost 200 K above its boiling point. Point out the extra vigour with which it boils when an object at room temperature is placed in it. This is comparable to putting a hot poker into water.

Oxygen is paramagnetic because it has two unpaired electrons while nitrogen does not.

Extensions

One teacher involved in the trials reports making a mercury hammer head by pouring mercury into a plastic mould and freezing a wooden handle into it. The hammer was used to drive a nail into a piece of wood.

THE ROYAL
SOCIETY OF
CHEMISTRY

Further details

Liquid nitrogen can be obtained from universities, higher education institutions, hospitals and industry. It is very inexpensive – about 10 p per dm^3 – and can be stored and transported in ordinary vacuum flasks. A 1.5 dm^3 vacuum flask full of liquid nitrogen will still be about 2/3 full after 24 hours. Polystyrene cool boxes can also be used to store the liquid for shorter periods of time. Vacuum flasks containing liquid nitrogen cannot be stoppered and a good way of transporting liquid nitrogen in a car is to use the arrangement in *Fig. 3* in which a small hole is made in the drinking cup/stopper of a vacuum flask and a plug of mineral wool is used to improve the insulation. Some means of keeping the flask upright in the car will be needed.

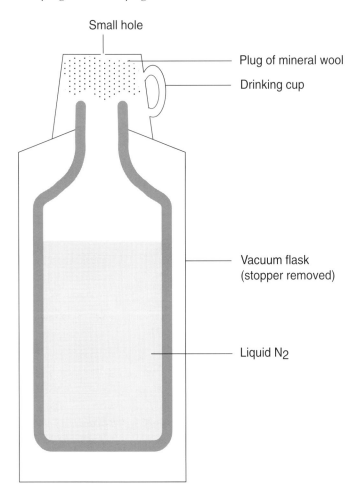

Fig. 3 Transporting liquid nitrogen

Safety

Wear eye protection and use insulating gloves when handling liquid nitrogen.

Care is needed when handling liquid oxygen and some teachers may prefer to omit this part of the experiment.

It is the responsibility of teachers doing this demonstration to carry out an appropriate risk assessment.

THE ROYAL
SOCIETY OF
CHEMISTRY

27. Demonstrating the colour changes of indicators using dry ice

Topic

Acids and bases, indicators, buffers. This is also an interesting 'fun' demonstration.

Timing

Up to 20 min.

Level

Introductory chemistry, but it could also be valuable for post-16 students.

Description

Dry ice is added to a measuring cylinder containing universal indicator (other indicators can also be used). Bubbles and a spectacular fog are produced and the indicator changes colour as the acidic carbon dioxide gas dissolves in the water.

Apparatus

▼   1 dm³ measuring cylinders – as many as the number of indicators to be used.

▼   Expanded polystyrene cool box to store the dry ice. The type of box in which Winchester bottles are often supplied is ideal.

Chemicals

The quantities given are for one demonstration.

▼   A few cm³ of universal indicator and/or other indicators as desired.

▼   A few cm³ of dilute **ammonia solution** and/or dilute **sodium hydroxide** solution.

▼   About 100 g of dry ice (solid carbon dioxide). This should be bought, since dry ice made from a carbon dioxide cylinder attachment will float and is less effective at saturating the solutions. Dry ice can be obtained from universities, higher education institutions, hospitals and industry. It can be stored and transported in expanded polystyrene boxes.

Method

The demonstration

Fill a 1 dm³ measuring cylinder to the 1 dm³ mark with water and add enough universal indicator to give an easily visible colour. Add a few drops of either ammonia solution or sodium hydroxide solution, to make the water alkaline, and stir. Add a few lumps of dry ice. These will sink to the bottom and bubble as CO_2 is given off. A spectacular fog will be produced at the top of the cylinder and, after several minutes, the indicator will change colour from blue to orange. The colour change will be more gradual if ammonia is used as the alkali as the reaction that occurs is a weak acid – weak base reaction. The final pH reached is about 4.5 and therefore, if colour changes that take place at pHs lower than this are to be shown, it will be necessary to add a little concentrated hydrochloric acid at the end.

THE ROYAL
SOCIETY OF
CHEMISTRY

Visual tips

Before adding the dry ice, pour a little of the coloured alkaline solution into a beaker and place this by the measuring cylinder to act as a reference colour.

Teaching tips

With the appropriate audience, this demonstration could be used to introduce a discussion of pH changes during the titrations of weak acids with strong and weak bases and hence buffers.

Because the colour changes take place gradually over about 15 minutes, it is a good idea to have something else to show the class in the meantime.

Theory

The relevant neutralisation equations are:

with ammonia
$$H_2O(l) + NH_3(aq) + CO_2(g) \rightarrow NH_4^+(aq) + HCO_3^-(aq)$$

with sodium hydroxide
$$OH^-(aq) + CO_2(g) \rightarrow HCO_3^-(aq)$$

Extensions

Other indicators can be used such as phenolphthalein (pink to colourless), thymolphthalein (blue to colourless), thymol blue (blue to yellow), phenol red (red to yellow) and bromothymol blue (blue to yellow).

Safety

Wear eye protection and use gloves to handle the dry ice as it can cause frostbite burns.

It is the responsibility of teachers doing this demonstration to carry out an appropriate risk assessment.

THE ROYAL
SOCIETY OF
CHEMISTRY

28. The alcohol 'gun'

Topic

Combustion and energy changes. The demonstration illustrates the principle of the internal combustion engine. It is also a spectacular 'fun' demonstration.

Timing

About 5 min.

Level

Pre-16.

Description

A plastic bottle is fitted with spark electrodes, filled with ethanol vapour and corked. The vapour is ignited with a spark and the cork is fired across the room with a small explosion.

Apparatus

▼ A polythene bottle of approximate capacity 500 cm³.

▼ Cork to fit the bottle.

▼ Rubber stopper, approximately size 17.

▼ Two paper clips.

▼ Two leads with crocodile clips on one end.

▼ EHT power pack (5 kV) or modified piezoelectric lighter (*see Fig. 2*).

▼ Retort stand with boss and clamp.

▼ G-clamp.

▼ Safety screen.

Chemicals

▼ A few cm³ of **ethanol**.

Method

Before the demonstration

Cut a hole in the base of the plastic bottle using a heated cork borer. The hole should be of such a size that a no. 17 rubber bung is a very tight fit. Make two holes about 5 mm apart in a no 17 rubber bung by straightening a paper clip, heating it in a Bunsen flame and pushing it through the bung. Insert a straightened paper clip through each hole so that each protrudes about 5 mm from the narrow side of the bung. These form the spark gap. Bend back the tails of the paper clips and push them into the side of the bung to stop them rotating (*Fig. 1*).

Connect the paper clips to an EHT power pack using leads and crocodile clips and adjust the size of the spark gap so that a large spark jumps when the voltage is turned up to about 4.5 kV. This adjustment is important if the 'gun' is to fire reliably. Alternatively, take a piezoelectric gas lighter and solder a pair of insulated leads onto the two terminals (*Fig. 2*). Each lead should have a crocodile clip on the end. Bend one of the terminals away from the other so that the lighter will not spark across the original terminals (*Fig. 2*). Connect the crocodile clips across the spark gap and adjust the gap to get a reliable spark on pressing the lighter.

Press the spark gap assembly firmly into the hole in the base of the bottle and check that sparking is still reliable.

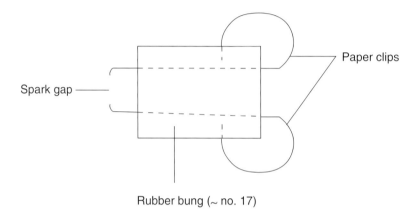

Fig. 1 The spark assembly

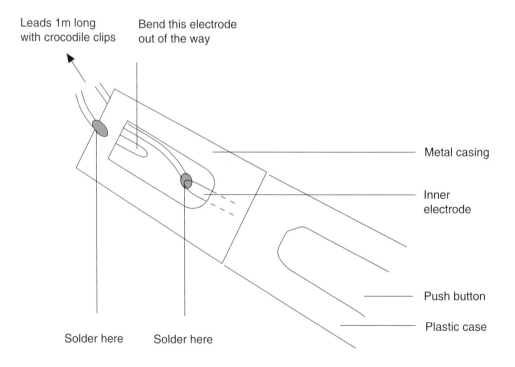

Fig. 2 Modifying a piezoelectric lighter

THE ROYAL
SOCIETY OF
CHEMISTRY

The demonstration

Clamp the plastic bottle so that the mouth is pointing safely away from the audience and any vulnerable equipment! *(Fig. 3)*. It will be necessary to attach the retort stand to the bench with a G-clamp. Squirt about 1 cm³ of ethanol into the bottle and shake to ensure vaporisation of the alcohol and mixing of the vapour with air. Place the cork gently in the mouth of the bottle, reclamp the bottle, and fire the 'gun' by turning up the EHT voltage or squeezing the piezoelectric lighter. It may be necessary to press the piezoelectric lighter several times before the mixture ignites. There will

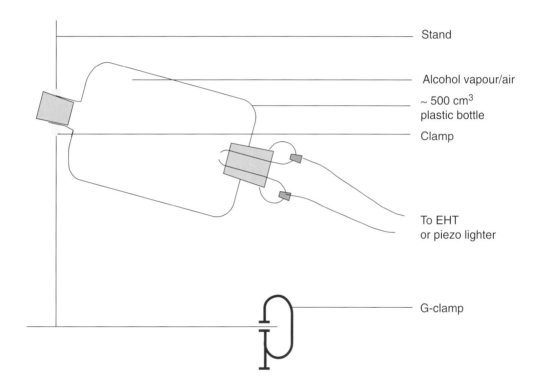

Fig. 3 The 'gun' about to fire

be a bang and the cork will be fired up to 5 m in the air. To fire again it will be necessary to replenish the air. This can be done by removing the sparking assembly and waving the bottle about, flushing the bottle with air from some type of bellows arrangement attached to a rubber tube or, ideally, by flushing the bottle with a compressed air line. It may be necessary to dry the sparking assembly with a tissue.

Visual tips

The flash of the explosion can be seen if the room is dark and is more easily seen if the bottle is transparent.

Teaching tips

The bang is not loud enough for students or the teacher to need to protect their ears. The experiment can be carried out without the cork in place and teachers may find it convenient to do this while practising the demonstration. Have a car spark plug available to show its similarity to the sparking assembly.

If there are problems getting reliable sparking, make sure that the spark gap assembly is dry.

Theory

The reaction (with ethanol) is:

$$C_2H_5OH(g) + 3O_2(g) \rightarrow 2CO_2(g) + 3H_2O(g) \quad \Delta H_c \ -1367 \ kJ \ mol^{-1}$$

Extensions

Bottles of smaller volumes could be used, but larger ones are not recommended.

Further details

The 'gun' can be fired with a Tesla coil if one is available. No leads are required, the Tesla coil is simply turned on and a spark applied to one of the paper clips.

A cork can be used instead of a rubber bung to construct the spark gap assembly. Alternatively, the spark gap of a piezoelectric lighter can be inserted directly into the hole in the bottle and glued in place. This does away with the need for leads, but leaves the operator closer to the explosion.

Safety

Wear eye protection. Consider where the cork could possibly go and have the audience wear eye protection. Use a safety screen. A cork will not carry as far as a rubber bung, and is less likely to cause damage. The cork could be tethered to the 'gun' with a length of string. Secure the apparatus to the bench.

It is the responsibility of teachers doing this demonstration to carry out an appropriate risk assessment.

29. The reaction between potassium permanganate and glycerol

Topic

Redox reactions. Exothermic reactions. This is also a spectacular 'fun' demonstration.

Timing

About 1 min.

Level

Any. Post-16 students may be able to appreciate the colours of the different oxidation states of manganese.

Description

Glycerol is poured onto potassium permanganate crystals. After a short lag time, steam is given off and a spectacular pink flame is produced.

Apparatus

▼   One clean tin lid – from a sweet tin for example.

▼   Heat proof mat.

Chemicals

The quantities given are for one demonstration.

▼   2–3 g of **potassium permanganate** (potassium manganate(VII)) in the form of fine crystals.

▼   About 1 cm³ of glycerol (propane-1,2,3-triol).

Method

The demonstration

Put 2–3 g potassium permanganate on a tin lid standing on a heat-proof mat and make a well in the centre of the pile. Pour about 1 cm³ of glycerol into the well. After about 20 seconds, steam is given off and a bright pink flame is produced which burns for a few more seconds.

Visual tips

A white background is useful. The reaction would look spectacular in a dark room.

Teaching tips

Dissolve the residue in water and a green solution will be seen (suggesting Mn(VI)) as well as a brown solid (suggesting Mn(IV)). This confirms the redox nature of the reaction. Point out the lilac flame, which is characteristic of potassium salts. Students may be able to suggest a plausible reason for the lag time; that the reaction is exothermic and at first slow, but that heat released by the initial reaction speeds up subsequent reaction and so on.

Theory

The potassium permanganate oxidises the glycerol to carbon dioxide and water (hence the steam) and is itself reduced.

Extensions

The reaction does not appear to work if crystals of potassium permanganate are sprinkled into glycerol.

Further details

Fine crystals of potassium permanganate appear to work much better than larger ones, so grind large crystals in a mortar and pestle. Some schools have reported that old glycerol is ineffective, possibly because it has absorbed water. One school reported that heating the glycerol to boiling and allowing it to cool before the demonstration alleviates this problem, perhaps by driving off water. Another school suggested drying the glycerol over calcium oxide.

Safety

Wear eye protection.

The residue from the reaction can be washed down the sink with plenty of water.

It is the responsibility of teachers doing this demonstration to carry out an appropriate risk assessment.

THE ROYAL
SOCIETY OF
CHEMISTRY

30. The non-burning £5 note

Topic

Combustion, but this is essentially a 'fun' demonstration.

Timing

About 2 min.

Level

Any.

Description

A piece of paper or a £5 note is soaked in a mixture of ethanol and water. The alcohol burns but the paper does not.

Apparatus

▼ Bunsen burner.

▼ A pair of tongs.

▼ A heat proof mat.

▼ Three 250 cm³ beakers.

Chemicals

The quantities given are for one demonstration.

▼ 75 cm³ of **ethanol**.

▼ A few grams of sodium chloride (common salt).

Method

Before the demonstration

Prepare some pieces of paper about the size of a £5 note.
Prepare three beakers – one containing about 50 cm³ of water; the second containing 50 cm³ of ethanol; and the third containing a mixture of 25 cm³ of water mixed with 25 cm³ of ethanol with a little sodium chloride dissolved in it.

The demonstration

Soak a piece of paper in water and try to ignite it by holding it with tongs in a yellow Bunsen flame. It will not ignite. Soak a second piece in ethanol. This will ignite easily. The alcohol will burn and ignite the paper, which will burn away. Soak a third piece in the alcohol-water mixture and hold it in the Bunsen flame. This time the alcohol will ignite and burn away, but the paper will not.

Visual tips

The pure alcohol flame is easily visible, but the alcohol-water one is almost invisible. The salt is added to colour the flame so that it can be seen. Some demonstrators may wish to explain this to the audience, others may prefer not to mention it. Alternatively the presence of the colourless flame may be shown by lighting a taper from it. The demonstration will look more impressive in subdued lighting.

THE ROYAL
SOCIETY OF
CHEMISTRY

Theory

The water in the alcohol-water mixture evaporates and keeps the temperature below the ignition temperature of paper (approximately 230 °C, but science fiction fans might remember this as 'Fahrenheit 451' – the temperature conversion could be an exercise for students). The paper will still be wet with water after the alcohol has burned away. Sodium chloride gives the flame the characteristic orange colour of sodium, which looks like a normal yellow flame.

Extensions

Different proportions of water and alcohol could be tried as could other alcohols.

Further details

The rich and/or confident may wish to try this with a £5 (or bigger!) note. The imaginative will be able to dream up some suitable patter to introduce the demonstration while the silver-tongued may be able to persuade a member of the audience to supply a note for 'burning'.

Safety

Wear eye protection.
    A fire extinguisher should be available.
    It is the responsibility of teachers doing this demonstration to carry out an appropriate risk assessment.

THE ROYAL
SOCIETY OF
CHEMISTRY

32. A giant silver mirror

Topic

The Tollens' test for aldehydes and reducing sugars. Also this is an excellent demonstration for general interest.

Timing

About 5 min.

Level

The Tollens' test is post-16, but the experiment will fascinate people of all ages.

Description

A solution of ammoniacal silver nitrate is reduced by glucose to silver, which forms a silver mirror on the inside of a large flask.

Apparatus

▼ One 1 dm³ flask with rubber stopper. A round bottomed flask looks most impressive, but any shape will do.

▼ One 250 cm³ beaker.

▼ Measuring cylinders – 25 cm³,100 cm³ and 250 cm³.

▼ Dropping pipette.

▼ Glass rod.

▼ Access to a fume cupboard (optional).

Chemicals

The quantities given are sufficient for three demonstrations.

▼ 8.5 g of **silver nitrate** ($AgNO_3$).

▼ 11.2 g of **potassium hydroxide** (KOH).

▼ 2.2 g of glucose (dextrose).

▼ 800 cm³ of deionised water.

▼ About 30 cm³ of **880 ammonia solution**.

▼ About 100 cm³ of **concentrated nitric acid**.

Method

Before the demonstration

Carefully clean the 1 dm³ flask. First use detergent and a brush, then rinse with water, followed by concentrated nitric acid and finally wash it out several times with deionised water. Thorough cleaning is vital if the demonstration is to succeed. Make up the three solutions as follows:

▼ Dissolve 8.5 g of silver nitrate in 500 cm³ of deionised water. This makes a 0.1 mol dm⁻³ solution.

THE ROYAL
SOCIETY OF
CHEMISTRY

▼   Dissolve 11.2 g of potassium hydroxide in 250 cm³ of deionised water. This
     makes a 0.8 mol dm⁻³ solution.

▼   Dissolve 2.2 g of glucose in 50 cm³ of deionised water.

The demonstration

Place 150 cm³ of the silver nitrate solution in a 250 cm³ beaker and, working in a
fume cupboard if possible, add 880 ammonia using a dropping pipette. A brown
precipitate will form. Continue to add the ammonia until the precipitate re-dissolves
to give a clear, colourless solution. Less than 5 cm³ of ammonia will be needed. The
solution then contains $Ag(NH_3)_2^+(aq)$.

Add 75 cm³ of the potassium hydroxide solution. A dark brown precipitate will
form. Add more ammonia dropwise until this precipitate redissolves to give a clear,
colourless solution. About 5 cm³ of ammonia will be needed.

Pour this solution into the 1 dm³ flask and add 12 cm³ of the glucose solution.
Stopper the flask and swirl the solution so that the whole of the inner surface of the
flask is wetted. The solution will turn brown. Continue swirling until a mirror forms.
This will take about 2 minutes.

When a satisfactory mirror has formed, pour the solution down the sink with
plenty of water. Rinse out the flask well with water and discard the washings down
the sink. The flask can now be passed around the class.

DO **NOT** SAVE THE SILVER SOLUTION IN A SILVER RESIDUE CONTAINER.

An alternative to plating the inside of a flask is to silver plate the outside of small
glass objects which can be suspended in the plating solution by hanging them on
threads. These objects must be cleaned beforehand.

Visual tips

The demonstration can be scaled up for greater impact or scaled down for economy.

Teaching tips

This reaction is the well known Tollens' or silver mirror test for aldehydes. The
method used to be used commercially for silvering mirrors.

Theory

Aldehydes such as glucose are reducing agents and will reduce $Ag^+(aq)$ ions to
metallic silver. They themselves are oxidised to carboxlyate ions. The reaction that
occurs is:

$$CH_2OH(CHOH)_4CHO(aq) + 2Ag(NH_3)_2^+(aq) + 3OH^-(aq) \rightarrow$$

$$2Ag(s) + CH_2OH(CHOH)_4CO_2^-(aq) + 4NH_3(aq) + 2H_2O(l)$$

Extensions

Try using an ordinary aldehyde instead of glucose, and show that the reaction does
not work with a ketone such as propanone.

Further details

The silver can be removed from the silvered flask with concentrated nitric acid. Work
in a fume cupboard because nitrogen dioxide is formed.

There are reports of silvered flasks being kept for several years as ornaments.

THE ROYAL
SOCIETY OF
CHEMISTRY

Safety

Wear eye protection.

There have been a few reports of alkaline ammoniacal silver nitrate solutions exploding after standing for some time. This rare occurrence is thought to be caused by the formation of silver nitride or silver fulminate. To avoid this risk, the ammoniacal silver nitrate solution should not be made up before the demonstration and any silvering solution left after the demonstration should not be placed in a silver residues container but should be washed down the sink with plenty of water. The silvered flask should be rinsed thoroughly with water and the washings washed down the sink as soon as the silvering has finished.

It is the responsibility of teachers doing this demonstration to carry out an appropriate risk assessment.

THE ROYAL
SOCIETY OF
CHEMISTRY

33. Determination of relative molecular masses by weighing gases

Topic

Gases, relative molecular masses.

Timing

About 10 min.

Level

Post-16.

Description

Known volumes of different gases are weighed in a gas syringe and their relative molecular masses are determined.

Apparatus

▼ One 50 cm³ (or larger, if available) plastic disposable syringe fitted with a hypodermic needle.

▼ One small rubber bung.

▼ Gas bags – one for each gas which is to be used (see demonstration No. 19).

▼ One nail about 5 cm long.

▼ Access to a balance – ideally three figure.

▼ Access to a barometer to measure atmospheric pressure (optional).

▼ Thermometer to measure room temperature.

Chemicals

▼ Access to sources of the required gases – either cylinders or chemical generators. **Hydrogen**, **oxygen**, nitrogen, **methane** (natural gas) and carbon dioxide would be a good selection.

Method

Before the demonstration
Fill the gas bags with the appropriate gases and label them.

Modify the syringe as follows:

Pull out the plunger so that the volume of air in the syringe is 50 cm³. Warm the nail in the Bunsen flame and push it through the stem of the plunger as shown in the figure. When the nail is in place, the plunger can be 'locked' at the 50 cm³ mark.

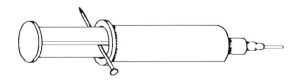

Locking the plunger at the 50 cm³ mark

THE ROYAL
SOCIETY OF
CHEMISTRY

The demonstration

With the syringe empty, stab the hypodermic needle into the rubber stopper to seal it. Withdraw the plunger and 'lock' it at the 50 cm³ mark with the nail. The syringe now contains 50 cm³ of vacuum. Weigh the syringe and note the mass. Now remove the rubber stopper from the needle and the nail from the hole and press in the syringe plunger. Inject the needle through the self-sealing stopper of one of the gasbags and suck 50 cm³ of gas into the syringe. Seal the gas in by sticking the needle into the rubber stopper and use the nail to 'lock' the syringe. It is not really necessary to lock the syringe in this case, but it does ensure that the nail is weighed along with the syringe. Weigh the syringe, stopper and nail. The difference between the two masses represents the mass of the gas under investigation. Flush out the gas and repeat this procedure with the other gases.

Calculate the relative molecular masses of the gases using $PV = nRT$ to calculate n, the number of moles of gas, which is equal to the mass of the gas in the syringe divided by its relative molecular mass in grams. Alternatively use the approximation that one mole of any gas occupies 24 dm³ under average room conditions. If the former method is used, to find out the atmospheric pressure use a barometer or ring the local meteorology service.

Visual tips

A balance that can be interfaced to a computer to provide an enlarged display on a monitor is useful.

Teaching tips

Many students will not understand why an empty syringe with the plunger on the zero mark cannot be used for the mass of the syringe with no gas. Be prepared to explain the buoyancy effect of displaced air. The weight of the syringe in this configuration can be measured and compared with its weight with the plunger locked at 50 cm³ to confirm that there is a difference.

Further details

With a 50 cm³ syringe it is difficult to determine the mass of hydrogen accurately unless a four figure balance is available.

Safety

Wear eye protection.

It is the responsibility of teachers doing this demonstration to carry out an appropriate risk assessment.

THE ROYAL
SOCIETY OF
CHEMISTRY

34. Flame colours

Topic

Flame colours of alkali and alkaline earth metals (and others).

Timing

About 5 min.

Level

Lower secondary but could also be useful to introduce a discussion of atomic spectra post-16. This is a spectacular demonstration for an audience of non-chemists.

Description

Solutions of alkali metal salts (and others) in ethanol are sprayed into a Bunsen flame and form spectacular jets of coloured flame.

Apparatus

▼ Trigger pump operated spray bottles such as those used for spraying house plants. Products such as window cleaner are sold in these bottles and empty ones can be cleaned with water and re-used for this experiment. Ideally, one bottle is needed for each metal, although it is possible to wash one out between solutions.

Chemicals

The quantities given are for one demonstration.

▼ About 10 cm^3 of **ethanol** for each metal.

▼ Less than 1 g of a salt of each metal. Chlorides are best, but other salts work and are worth trying if chlorides are not available. Examples of compounds that work well are: sodium chloride, potassium chloride, lithium iodide and **copper sulphate**.

Method

Before the demonstration
Make a saturated solution of each salt in about 10 cm^3 of ethanol. Only a few mg of each is required. Place each solution in a spray bottle and label it.

The demonstration
Adjust the nozzles of the spray bottles to give a fine mist and spray the solutions into a roaring Bunsen flame. Take care to direct the spray away from yourself and the audience. The colour of the resulting jet depends on the metal used. The solutions can be retained for future use and can be stored in the plastic bottles for several weeks, at least, without apparent deterioration of the bottles.

Visual tips

The demonstration looks most spectacular in a darkened room.

Teaching tips

Students can observe line spectra through hand held spectroscopes.

THE ROYAL
SOCIETY OF
CHEMISTRY

Extensions

A solution of boric acid in ethanol (1 g in 10 cm^3 makes a saturated solution) gives a green flame. This is used as the basis of a qualitative test for borates.

Further details

Ensure that the spray bottles are trigger operated, with a piston rather than a scent spray pump where a rubber bulb is squeezed. This will prevent any possibility of the flame flashing back into the container.

Safety

Wear eye protection.

It is the responsibility of teachers doing this demonstration to carry out an appropriate risk assessment.

THE ROYAL
SOCIETY OF
CHEMISTRY

35. The hydrogen rocket

Topic

Combustion, energy changes in reactions. This is an excellent 'general interest' demonstration.

Timing

About 5 min.

Level

Any.

Description

A plastic pop bottle 'rocket' is filled with hydrogen and air. The gas mixture is ignited with an electric spark and the 'rocket' will fly several metres. The rocket can be set up to fly along a wire or launched from a short section of drainpipe.

Apparatus

▼　One 1 dm^3 (or 500 cm^3) plastic fizzy drink bottle.

▼　Two rubber bungs to fit the bottle.

▼　Plastic washing up bowl to act as a pneumatic trough for filling the bottle with hydrogen.

▼　Modified piezoelectric gas lighter *(Fig. 2)* or EHT (0–5kV) power pack with two leads each about 1 m long and crocodile clips.

▼　Two paper clips.

▼　Safety screen.

For the 'flight on wire' method:

▼　One piece of softwood about 15 cm x 2 cm x 2 cm.

▼　Two small screw eyes *(Fig. 3)*.

▼　Two strong rubber bands to fit tightly around the circumference of the pop bottle.

▼　String – as long as the demonstration room (perhaps 10 m).

For the drainpipe launch method:

▼　About 30 cm of plastic drainpipe into which the plastic bottle will slide.

Chemicals

▼　Access to a **hydrogen** cylinder with a regulator and a length of rubber tube.

Method

Before the demonstration

Construct a sparking assembly as follows. Make two holes about 5 mm apart in one of the rubber bungs by straightening a paper clip, heating it in a Bunsen flame and

THE ROYAL
SOCIETY OF
CHEMISTRY

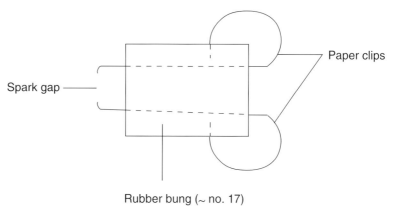

Spark gap

Paper clips

Rubber bung (~ no. 17)

Fig. 1 The spark assembly

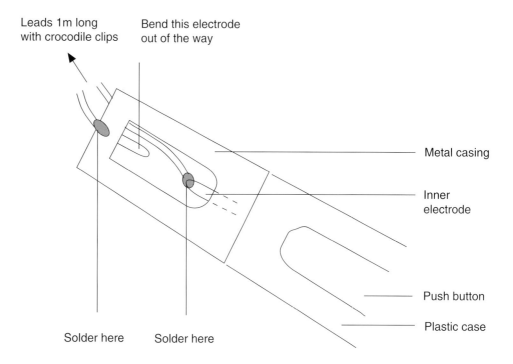

Leads 1m long
with crocodile clips

Bend this electrode
out of the way

Metal casing

Inner
electrode

Push button

Plastic case

Solder here   Solder here

Fig. 2 Modifying a piezoelectric lighter

pushing it through the bung. Insert a straightened paper clip through each hole so that each protrudes about 5 mm from the narrow side of the bung. These form the spark gap. Bend back the tails of the paper clips and push them into the side of the bung to stop them rotating *(Fig. 1)*.

Construct a spark generator as follows:

Take a piezoelectric gas lighter (which can be bought from a hardware shop) and solder a pair of insulated leads about 1 m long onto the two terminals. Each lead should have a crocodile clip on the end. Bend one of the terminals away from the other so that the lighter will not spark across the original terminals *(Fig. 2)*. Connect the crocodile clips across the spark gap assembly and adjust the gap to get a reliable spark on pressing the lighter.

Press the spark gap assembly firmly into the mouth of the bottle and check that sparking is still reliable.

Alternatively, connect an EHT supply across the sparking assembly and check that turning up the EHT voltage will produce a spark.

For the 'flight on wire' method:

Screw the screw eyes into the piece of wood about 2 cm from each end. Thread the string through the screw eyes and run the string across the laboratory at a suitable height to avoid obstructions, attaching it firmly at both ends (*Fig. 3*). One end must be able to allow the spark assembly to be clamped close to the string. If the string slopes gently upwards, it will help to slow the rocket down.

For the drainpipe launch method:

Clamp the drainpipe 'launch tube' securely so that the rocket will be launched safely away from the audience and any obstructions. Bear in mind that it will fly several metres. It is preferable to do the launch outside. It will be necessary to clamp the spark assembly at the base of the launch tube.

The demonstration

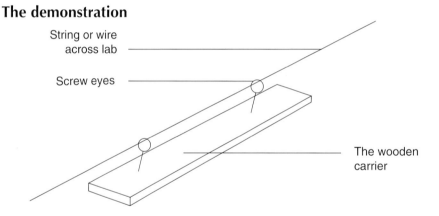

String or wire across lab

Screw eyes

The wooden carrier

Fig. 3 String and screw eyes assembly for supporting the rocket

Fill the plastic bottle with water, invert it over a bowl full of water and fill it about 2/7 full of hydrogen (2:5 is the stoichiometric ratio for hydrogen and air by volume). It may help to have previously marked the bottle at the correct level with a marker pen. Keeping the bottle upright and mouth down, lift it from the water and allow air to replace the remaining water. Cork the bottle and carry it to the launch point. Replace the cork with the spark assembly, clamp the spark assembly and either attach the rocket firmly to the wooden carrier with rubber bands or place the rocket in the launch tube. Note that the spark assembly must be clamped firmly (*Fig. 4.*). If a retort stand is used, it will need to be attached to the bench with a G-clamp.

Fire the rocket by operating the sparking device.

THE ROYAL
SOCIETY OF
CHEMISTRY

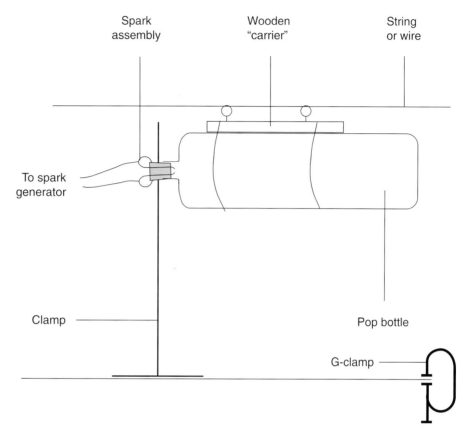

Fig. 4 Hydrogen rocket before take-off

Teaching tips

Point out that the inside of the rocket is covered with condensation after firing. The demonstration could lead to a discussion of the potential of hydrogen as a fuel and a consideration of a hydrogen economy.

Extensions

Try different ratios of hydrogen to air. Methane can be tried instead of hydrogen by adjusting the air to gas ratio to 1 volume of methane to 10 volumes of air but is disappointing.

Safety

Wear eye protection.

Set up a safety screen around the firing point. Secure the apparatus to the bench.

Wear ear protection and insist that members of the audience put their fingers in their ears.

It seems sensible to warn others within earshot that loud bangs are likely.

It is the responsibility of teachers doing this demonstration to carry out an appropriate risk assessment.

THE ROYAL
SOCIETY OF
CHEMISTRY

36. A controlled hydrogen explosion

Topic

Combustion, energy changes in reactions. This is an excellent general interest demonstration.

Timing

About 5 min.

Level

Any. It can be used to introduce a discussion of bond energies to post-16 or possibly pre-16 students.

Description

A coffee tin is filled with hydrogen via a hole in its base. The hydrogen is burnt at a second hole in the lid. Eventually an explosive hydrogen-air mixture is formed and the lid of the can is blown off with a loud bang.

Apparatus

▼   Tripod.

▼   500 g catering size coffee tin with press on lid.

▼   Safety screen.

▼   Ear protectors for the demonstrator (the CDT department may have some).

Chemicals

▼   Access to a **hydrogen** cylinder with a regulator and a length of rubber tubing.

Method

Before the demonstration
Make a small pencil sized, *ie* approximately 4 mm diameter, hole in the lid of the coffee tin. Make a second, approximately 1 cm diameter hole in the base of the tin.

The demonstration
Put the lid on the tin and stand the tin, lid uppermost, on a tripod. Place a safety screen between the tin and the audience. Insert the tube from the hydrogen cylinder into the hole in the base of the tin and fill the tin with hydrogen. Allow at least a minute to ensure that the tin is full of hydrogen and then turn off the hydrogen supply and remove the tube. Now light the hydrogen at the hole in the lid and stand back (*see Fig*). The hydrogen will burn first with a yellow and then a blue flame. This will gradually get smaller. Air is drawn into the tin through the bottom hole and after about thirty seconds, an explosive mixture results. There will be a loud bang and the lid will fly off the tin.

Teaching tips

Point out that this is an exothermic reaction and that energy is produced in the forms of heat, light, sound and kinetic (the flying lid).

$$2H_2(g) + O_2(g) \rightarrow 2H_2O(g) \quad \Delta H = -484 \text{ kJ mol}^{-1}$$

THE ROYAL
SOCIETY OF
CHEMISTRY

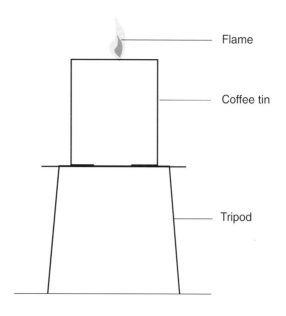

Flame

Coffee tin

Tripod

Exploding tin

The transformation of chemical energy into kinetic energy is comparable with the process occurring in an internal combustion engine.

A suitable group could be given bond energy calculations to do.

Theory

Hydrogen and air is explosive when the mixture contains between 4 % H_2 and 75 % H_2. The flame burns quietly until the mixture falls into this range.

Extensions

Natural gas (methane) may be used instead of hydrogen. The explosion limits for methane and air are 5 % –15 % of methane.

Further details

Cardboard Pringles® potato crisps boxes have been suggested as an alternative to the coffee tin.

Safety

Wear ear protection and insist that members of the audience put their fingers in their ears.

Use a safety screen between the tin and the audience.

Some teachers may prefer to light the hydrogen using a candle taped to the end of a metre ruler.

Hydrogen is extremely flammable and forms explosive mixtures with air.

It is not recommended to attempt this demonstration using hydrogen from a chemical generator because of the difficulty of generating the gas fast enough to reliably sweep air out of the apparatus.

Take care not to site the tin below a light fitting as the lid may well hit the roof.

Warn others within earshot that loud bangs are likely.

It is the responsibility of teachers doing this demonstration to carry out an appropriate risk assessment.

THE ROYAL
SOCIETY OF
CHEMISTRY

37. Exploding balloons

Topic

Combustion, exothermic reactions. This is an excellent general interest demonstration.

Timing

About 5 min.

Level

Any, but it could be used to start a discussion on bond energies for an appropriate group of students.

Description

Party balloons containing a variety of gases and gas mixtures are ignited and the vigour of the resulting reactions is noted.

Apparatus

▼   Four party balloons plus more for repeats.

▼   Metre rule, candle and insulating tape.

▼   Ear protectors for the demonstrator (the CDT department may have some).

▼   Cotton thread.

Chemicals

▼   Access to cylinders (with regulators and rubber delivery tubing) of **hydrogen**, **oxygen**, helium or 'balloon gas' (helium + a little air) (optional). The latter is available from BOC, but is optional given its expense – costing about £30–£40 (1994) for 50–200 balloons-worth of gas including cylinder hire.

Gases generated chemically do not have sufficient pressure to inflate balloons.

Method

Before the demonstration
Attach a candle to the end of a metre rule with insulating tape.

The demonstration
Inflate a balloon with air by mouth. Touch the lighted candle to it and it will burst with a small pop. This is due solely to the rubber bursting. Alternatively inflate a balloon with helium or balloon gas (if available) and burst it in the same way. This method has the advantage that the balloon will float as will all the others. Inflate a second balloon with hydrogen from a cylinder. This will float, so tether it to the bench with a length of thread. Apply the lighted candle on the metre rule at arm's length. The balloon will explode with a loud bang and some flames due to the reaction of hydrogen with atmospheric oxygen. Place a little hydrogen into a third balloon and complete the inflation with air from the mouth. The stoichiometric ratio of hydrogen to air is 2:5, but demonstrators might prefer to have rather more hydrogen to ensure the balloon will float. Ignite the mixture as before. The bang will be louder this time because the air and hydrogen are better mixed. Finally, inflate the third balloon. Use a little hydrogen from the cylinder and top up with oxygen from a cylinder. The ideal mixture is two volumes of hydrogen to one of oxygen but

THE ROYAL
SOCIETY OF
CHEMISTRY

this may need adjusting if you want the balloon to float. Wear ear protection and ensure that members of the audience place their fingers in their ears. Ignite the balloon keeping as far away from the balloon as possible. There will be a very loud bang.

Visual tips

If the experiment is done in a dark room, it is more spectacular and the flames are easier to see.

Teaching tips

If the balloons are first inflated by mouth and then deflated, they are easier to inflate later.

A good way of presenting the demonstration is to have the balloons, all floating and tethered to the bench, prepared shortly before the demonstration and then to ignite them one by one in order of increasing vigour of reaction.

Ask the audience why helium is now used in airships rather than hydrogen. What advantage does hydrogen have?

Theory

Hydrogen – air mixtures will explode between 4% and 75 % H_2.

$$2H_2(g) + O_2(g) \rightarrow 2H_2O(g) \qquad \Delta H = -484 \text{ kJ mol}^{-1}$$

Extensions

Try varying the proportions of hydrogen and air (or oxygen).

Hydrogen from a cylinder can be used to blow bubbles in a bubble mixture. These can be ignited with a candle on a metre rule as they float upwards. One teacher reported that those ignited just after being blown burn softly while those ignited near the ceiling 'pop' loudly. This is presumably due to the diffusion of oxygen into the bubble.

Safety

Wear eye protection.

Ear protection is essential for the hydrogen-oxygen explosion and desirable for the others.

Warn others within earshot that loud bangs are likely.

It is the responsibility of teachers doing this demonstration to carry out an appropriate risk assessment.

38. The combustion of methane

Topic

Combustion, exothermic reactions. This is an excellent general interest demonstration.

Timing

Less than 5 min.

Level

Any, but it could be used along with an introduction to the Bunsen burner.

Description

A large coffee tin is fitted with a glass chimney and filled with methane. The gas is lit at the top of the chimney, burns down the chimney and eventually explodes inside the tin, blowing the lid off.

Apparatus

▼   One 500 g or 750 g catering size coffee tin with press on lid.

▼   About 50 cm of glass tubing, roughly 2–3 cm in diameter.

▼   Araldite or other epoxy resin adhesive.

▼   A length of rubber tube to reach from the gas tap to the apparatus.

▼   Safety screen.

Chemicals

▼   **Methane** (natural gas) from the gas tap.

Method

Before the demonstration

Make a hole about 1 cm in diameter in the base of the coffee tin. Make a larger hole to take the glass tubing half way up the side of the tin. Use Araldite to glue the glass tube in place (*see Fig*).

The demonstration

Clamp the tin so that the chimney is vertical and the lid is pointing away from the audience. Place a safety screen between the tin and the audience. Using a length of rubber tube attached to the gas tap and placed through the hole in the base of the tin, fill the tin with methane gas. Allow at least a minute to ensure that all the air is swept out of the apparatus. Remove the rubber tube and turn off the gas. Light the gas at the top of the chimney. It will initially burn with a yellow luminous flame. This will change to a blue flame as more air is drawn up the chimney. Eventually the blue flame will descend the chimney. As it reaches the bottom, the gas in the tin explodes and the lid is blown off. The explosion is fairly gentle and the lid must not be on too tightly or it will not be blown off.

Visual tips

Perform the experiment in a dark room for best visibility.

THE ROYAL
SOCIETY OF
CHEMISTRY

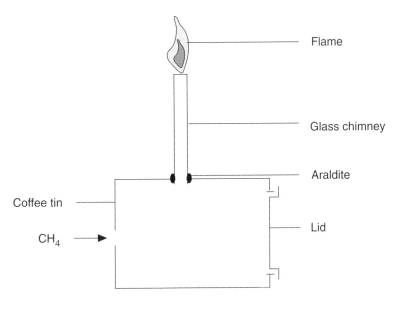

Flame

Glass chimney

Araldite

Coffee tin

Lid

CH_4 →

Combustion of methane

Teaching tips

Point out the similarity of the flame to the flames of a Bunsen burner with the air hole opened and closed.

Theory

Mixtures of methane and air with between 5% and 15 % methane will explode. The flame descends the chimney because the reaction is using up gas faster than the gas can rise up the chimney.

Safety

Wear eye protection.

It is not recommended to attempt this experiment with hydrogen: the explosion is both rapid and violent.

It is the responsibility of teachers doing this demonstration to carry out an appropriate risk assessment.

THE ROYAL
SOCIETY OF
CHEMISTRY

39. Equilibria involving carbon dioxide and their effect on the acidity of soda water

Topic

Equilibria, Le Chatelier's principle, solubility of gases (Henry's Law).

Timing

About 5 min.

Level

Post-16.

Description

Soda water is placed in a syringe and the plunger pulled out to reduce the pressure above it. Carbon dioxide is seen out-gassing and methyl red indicator turns from red to yellow, showing that the solution has become less acidic.

Apparatus

▼   One 50 cm³ plastic syringe. (A smaller one will work but is less easily visible.)

▼   Syringe cap (optional).

▼   One 5 cm nail.

▼   One small beaker.

Chemicals

▼   A few drops of methyl red indicator. To make the stock solution, dissolve 0.02 g of solid methyl red in 60 cm³ of ethanol and make up to 100 cm³ with distilled water.

▼   A few cm³ of soda water or carbonated mineral water. A fresh, unopened bottle is best. Lemonade or tonic water do not work because they contain citric acid.

Method

Before the demonstration
Modify the syringe as follows:
    Pull out the plunger so that the volume of air in the syringe is 50 cm³ *(see Fig)*. Warm the nail in a Bunsen flame and push it through the stem of the plunger as shown in the diagram. When the nail is in place, the plunger can be 'locked' at the 50 cm³ mark.

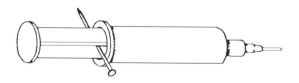

Locking the plunger at the 50 cm³ mark

THE ROYAL
SOCIETY OF
CHEMISTRY

The demonstration

Pour a few cm^3 of soda water into the beaker and add a few drops of methyl red to give a red solution. Draw about 5 cm^3 of this solution into the syringe. Place a syringe cap over the end of the syringe (or use a finger), pull the plunger out to the 50 cm^3 mark and lock it with the nail. Bubbles of carbon dioxide will be seen out-gassing and the indicator will begin to turn orange. Shake the syringe to speed up the out-gassing. Hold the syringe vertically with the nozzle pointing upwards, remove the syringe cap and the nail, and push in the plunger to expel the gas but not the solution. Stopper the syringe again and repeat the above cycle. More CO_2 bubbles will be seen and the indicator will turn more towards a yellow colour. Several more cycles can be repeated until the indicator becomes yellow.

Visual tips

A white background helps. Place the syringe next to the original red solution to emphasise the colour change.

Teaching tips

An assistant may be needed to help lock the syringe with the nail.

Students may not be familiar with methyl red. If so, demonstrate its colours in acid and alkali beforehand. It is red below pH 4.2 and yellow above pH 6.3.

Boil some soda water containing a little methyl red. This will expel the carbon dioxide, which is less soluble at high temperatures, and also show the colour change of the indicator from red to yellow.

Theory

Soda water contains carbon dioxide that has been dissolved in it under pressure (Henry's Law). The equilibria involved are:

$$CO_2(g) \rightleftharpoons CO_2(aq) \qquad\qquad (1)$$

$$CO_2(aq) + H_2O(l) \rightleftharpoons H_2CO_3(aq) \qquad\qquad (2)$$
$$\text{carbonic acid}$$

$$H_2CO_3(aq) \rightleftharpoons H^+(aq) + HCO_3^-(aq) \qquad\qquad (3)$$
$$\text{hydrogencarbonate ions}$$

$$HCO_3^-(aq) \rightleftharpoons H^+(aq) + CO_3^{2-}(aq) \qquad\qquad (4)$$
$$\text{carbonate ions}$$

Thus the solution is acidic.

Reducing the pressure allows CO_2 to come out of solution *ie* drags equilibrium (1) to the left. The result is that the other three equilibria also move to the left, removing $H^+(aq)$ ions from the solution and making the solution less acidic.

For simplicity, some teachers may prefer not to discuss equilibrium (4).

Safety

Wear eye protection.

It is the responsibility of teachers doing this demonstration to carry out an appropriate risk assessment.

THE ROYAL
SOCIETY OF
CHEMISTRY

40. Thermal properties of water

Topic

Properties of materials: specific heat capacity, boiling point and thermal conductivity and the effects of hydrogen bonding in water.

Timing

About 5 min.

Level

Lower school, but it could be used with post-16 students to illustrate the effect of hydrogen bonding on the properties of water.

Description

Water can be boiled in a paper cup without burning the cup. A balloon containing water does not burst when a lighter flame is applied to it.

Apparatus

▼   Bunsen burner.

▼   Tongs.

▼   Heat proof mat.

▼   Cigarette lighter.

▼   Two disposable waxed paper cups (as used for cold drinks at parties). Alternatively, paper cake cases can be used.

▼   Two party balloons.

▼   Thermocouple-type thermometer with a large display *eg* interfaced to a computer and monitor (optional).

Chemicals

▼   Water.

Method

The demonstration
Method 1
Using tongs, hold a paper cup over a fairly small Bunsen flame (gas about half on, air hole about half open). The cup will catch fire within a few seconds. Allow it to burn out or extinguish it with the heat proof mat. Half fill the second paper cup with tap water and hold it over the same flame. Take care to centre the flame on the base of the cup and ensure that the flame does not play on the sides of the cup above the water level. After a few minutes the water will boil and the cup will remain undamaged except for a little charring around the rim on the base.

Method 2
Inflate one of the balloons by mouth to the usual size and knot the end. Put about 100 cm³ of water from the tap in the second balloon and then inflate to the same size as the first balloon and tie the neck. Hold or clamp the first balloon and apply a cigarette lighter flame to its base. It will burst almost instantly. Hold or clamp the

THE ROYAL
SOCIETY OF
CHEMISTRY

second balloon similarly and apply the lighter flame to the base where the water is. The balloon will not burst and the flame can be held in place for some time. If desired show that this is caused by the presence of the water by moving the flame to a part of the balloon not filled with water. It will burst instantly, so work over the sink or a tray.

Visual tips

Arrange a suitable background so that steam can be clearly seen coming from the cup. Black is better than white. If available, a display thermometer can be placed in the cup to monitor the temperature changes.

Theory

Paper will not ignite below about 230 °C (science fiction fans might remember Fahrenheit 451!). The specific heat capacity (SHC) of water is high (4.2 J g^{-1} K^{-1}) and so it takes a lot of heat to produce a relatively small temperature rise in water. The same amount of heat would produce a much bigger temperature rise in paper. The high SHC of water is due to the strong intermolecular hydrogen bonds – it takes a lot of energy to separate water molecules. In any case the temperature of the water cannot rise above its boiling point of 100 °C. Water is a relatively good conductor of heat and convection effects also transfer heat away from the hot spot above the Bunsen flame.

In the absence of water, the rubber of the balloon soon heats up and softens and the balloon bursts. The high SHC of water prevents this when the balloon is filled.

Safety

Wear eye protection.

It is the responsibility of teachers doing this demonstration to carry out an appropriate risk assessment.

THE ROYAL
SOCIETY OF
CHEMISTRY

41. The density of ice

Topic

Water, hydrogen bonding, but it is also an interesting exercise in observation skills.

Timing

About 10 min.

Level

Lower secondary or post-16 to introduce work on hydrogen bonding and the structure of ice.

Description

Ice cubes float on cooking oil but, on melting, the water that is produced sinks.

Apparatus

▼   One 1 dm^3 measuring cylinder.

Chemicals

The quantities given are for one demonstration.

▼   Water.

▼   A few ice cubes. These may be made with a little added food colouring (blue is good) for better visibility.

▼   400 cm^3 of cooking oil *eg* Tesco's pure vegetable oil.

Method

Before the demonstration

Make the ice cubes with a few drops of food colouring per cube. Ensure that they are completely frozen. Partly frozen cubes may have liquid water trapped inside which will affect their density.

     Check that ice cubes do actually float on the brand of cooking oil to be used.

The demonstration

Place about 400 cm^3 of water and 400 cm^3 of cooking oil in the measuring cylinder. Allow the two layers to separate fully; the oil will be on the top. Drop an ice cube into the cylinder. It will float (just) on top of the oil. Watch the cube. As it melts, the water that is formed makes a droplet attached to the cube. Eventually this detaches itself from the cube and sinks, joining the water layer below. This illustrates the anomalously greater density of water compared to ice.

     A number of other interesting observations can be made:

▼   After most of the cube has melted the weight of the water drop is sufficient to drag the ice cube down with it *ie* the average density of the cube and drop is greater than that of the oil. Sometimes, as the cube and drop are sinking, the drop detaches itself from the cube and the cube floats back to the surface.

▼   Small mini droplets occasionally break off from the main one as it descends forming a 'string of pearls' effect.

▼   Water droplets may sit for some time on the water oil interface without coalescing with the body of the water.

THE ROYAL
SOCIETY OF
CHEMISTRY

▼ When the coloured water droplets begin to coalesce with the water, the coloured water can be seen to sink as it mixes, because of its greater density.

▼ There are interesting changes of shape in the water droplet as it forms, detaches from the cube and as it sinks.

Visual tips

Water that has been dyed blue is easily seen in the pale yellow oil against a white background.

Teaching tips

Repeat the demonstration with an uncoloured ice cube to show that the colouring makes no difference.

Theory

The density of ice is about 0.92 g cm^{-3} and that of water is about 1.00 g cm^{-3} at 0 °C (*Fig. 1*). Cooking oil has a density between these two and therefore ice floats on the oil whereas water sinks. Most solids are denser than their liquids. The lower density of ice is caused by its structure, a hydrogen bonded tetrahedral network similar to that of diamond (*Fig. 2*).

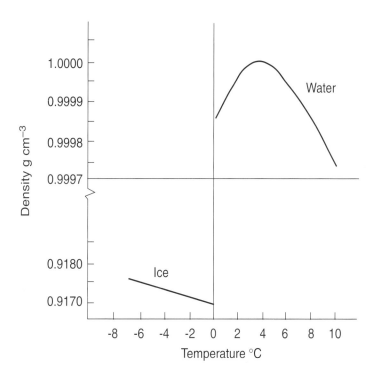

Fig.1 Temperature dependence of the density of ice and water

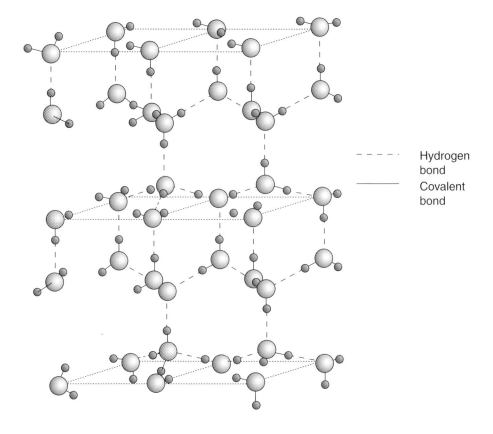

Hydrogen
bond
Covalent
bond

Fig. 2 Hydrogen bonded structure of ice

Safety

Wear eye protection.

It is the responsibility of teachers doing this demonstration to carry out an appropriate risk assessment.

Acknowledgement

This demonstration is based on an idea developed by Colin Johnson at Techniquest, Cardiff.

42. The tubeless siphon

Topic

This experiment will probably be done for general interest/entertainment but it is a spectacular demonstration of hydrogen bonding.

Timing

About 5 min.

Level

Any, for entertainment, but post-16 if it is used to illustrate hydrogen bonding.

Description

A viscous liquid is poured from one beaker to another. When the first beaker is returned to the upright position, the liquid continues to siphon by climbing up the wall of the beaker and over the rim.

Apparatus

▼   One 1 dm³ beaker.

▼   One 2 dm³ beaker.

▼   One 250 cm³ beaker.

▼   A 30 cm ruler to act as a wide-bladed stirrer.

▼   A magnetic stirrer with a large follower.

Chemicals

The quantities given are for one demonstration.

▼   11 g of polyethylene oxide of relative molecular mass about 4×10^6. This compound can be obtained from Aldrich.

▼   100 cm³ of propan-2-ol (isopropanol).

▼   A little flourescein or food dye (optional).

Method

Before the demonstration

Pour 1 dm³ of water into the 2 dm³ beaker. Add a little fluorescein or food dye to give the desired colour if required. Weigh 11 g of polyethylene oxide into the 250 cm³ beaker and make this into a thin suspension with about 70 cm³ of the propan-2-ol. (Grinding the polyethylene oxide into a fine powder in a mortar and pestle speeds up the dissolution but is not essential.) Stir the water in the 2 dm³ beaker vigorously with the ruler and add the suspension slowly, avoiding the formation of lumps. It is helpful to have an assistant to pour in the slurry while the mixture is being stirred. Wash any remaining polyethylene oxide into the large beaker with the rest of the propan-2-ol. Continue to stir vigorously for several minutes until the mixture has the consistency of wallpaper paste then continue stirring with a magnetic stirrer for a few hours. Leave the mixture for about a day to hydrate fully and then pour it into the 1 dm³ beaker.

The demonstration

Place the 2 dm³ beaker on the bench. Tilt the beaker containing the mixture so that a little of it pours into the larger beaker, and then return the small beaker to almost an upright position. The mixture will continue to siphon out of the small beaker and into the large one *(see Fig)*. The siphon can be stopped by cutting the liquid with scissors close to the upper beaker. The upper liquid column will retract into the upper beaker. Siphoning eventually stops when the head, h, becomes too great. This usually occurs when about half of the liquid has been siphoned.

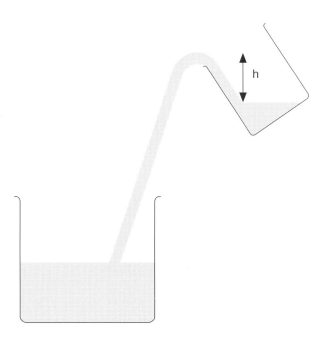

The tubeless siphon

Visual tips

Coloured liquid is seen most easily against a white background.

Theory

The solution is a so-called non-Newtonian fluid. The easiest way to explain the phenomenon to students is in terms of the large amount of hydrogen bonding between the long chain polyethylene oxide molecules and water molecules.

Extensions

This experiment lends itself to investigations. For example, what range of concentrations allows tubeless siphoning? Does changing the temperature have any effect? What is the maximum head that can be obtained? Can the viscosity of the mixture be measured? Does siphoning occur with polymers of shorter chain length (a range of relative molecular masses can be bought)? Does the time the solution is left after being made up have any effect?

Further details

D. F. James, *Nature (London)*, 1966, **212**, 754 is one of the first reports of this phenomenon.

THE ROYAL
SOCIETY OF
CHEMISTRY

One teacher reported that the larger the distance between the upper and lower beakers, the more mixture pulled out of the upper beaker.

Polyethylene oxide of relative molecular mass 900 000 does not appear to siphon in the same way.

Safety

Wear eye protection.

The solution can be disposed of down the sink with plenty of water. Alternatively, teachers may wish to keep the solution for further use, especially as polyethylene oxide is quite expensive. The solution should remain usable for some months when kept in a stoppered jar. A few drops of 2 % thymol solution in ethanol can be added to prevent bacterial growth.

It is the responsibility of teachers doing this demonstration to carry out an appropriate risk assessment.

THE ROYAL
SOCIETY OF
CHEMISTRY

43. Movement of ions during electrolysis

Topic

Electrolysis. There are also links with physics (the motor effect) and an analogy can be drawn with mass spectrometry.

Timing

About 10 min.

Level

14–15 year olds, but it could be shown to post-16 students to illustrate the deflection of ions in a mass spectrometer.

Description

A petri dish containing sodium sulphate solution and bromocresol purple indicator (yellow in acid, blue in alkali) is placed on an overhead projector. The solution is electrolysed and a blue colour streaming from the cathode shows the movement of hydroxide ions. These can then be deflected by a magnetic field and shown to obey the left hand motor rule.

Apparatus

▼   One petri dish – preferably glass because this is more transparent than plastic.

▼   0–12 V low voltage power pack.

▼   Two leads with crocodile clips.

▼   A flat 'magnadur' type magnet.

▼   A ring-shaped ceramic magnet such as one recovered from an old loudspeaker (optional). Ideally the ring magnet should be the same external diameter as the petri dish, but smaller ones can be used.

▼   An overhead projector and screen.

Chemicals

The quantities given are for one demonstration.

▼   7 g of anhydrous sodium sulphate.

▼   A few drops of bromocresol purple indicator solution (approximately 1 % solution in ethanol, note that this is more concentrated than the normal indicator solution).

▼   A few cm^2 of aluminium foil.

▼   A little dilute acid.

Method

Before the demonstration

Dissolve about 7 g of sodium sulphate in 100 cm^3 of water. Add several drops of bromocresol purple solution to give a deep blue colour. Add the minimum amount (no more than a drop or two) of dilute acid just to turn the indicator yellow. Make sure you know the polarities of the faces of the magnets.

THE ROYAL
SOCIETY OF
CHEMISTRY

The demonstration

Stand the petri dish on the centre of the stage and focus the OHP. Clip the crocodile clips to the rim of the petri dish on opposite sides. They should dip into the solution forming electrodes. Connect the crocodile clips to the 12 V DC terminals of the power pack and switch on. Bubbles of hydrogen will be seen at the cathode and a blue colour will spread from the cathode. This is caused by $OH^-(aq)$ ions which remain in the solution and are repelled from the cathode after the discharge of $H^+(aq)$.

Now hold one of the flat pole faces of the magnet above the petri dish, pole face horizontal, so that the magnetic field passes vertically through the petri dish. Take care not to obscure the audience's view – tweezers are useful. Observe the movement of the coloured stream to the left or right (depending on the direction of the magnetic field). Turn the magnet over and show that the stream now moves in the opposite direction (*Fig. 1*). Confirm that the deflection is in the direction predicted by the left hand motor rule remembering that the ions are negatively charged so that, conventionally, current flows in the opposite direction to that in which the ions are moving.

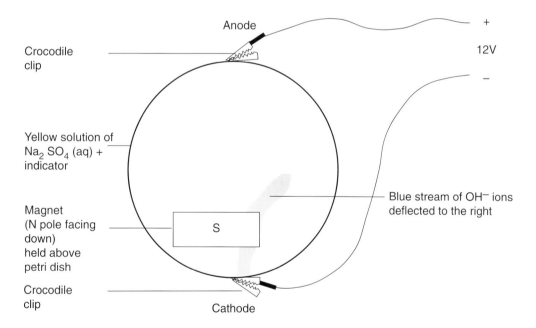

Fig. 1 Movement of ions in a magnetic field

Alternatively, stand the petri dish over a ring magnet so that the cathode can be seen through the centre of the ring (*Fig. 2*). The deflection can be seen more easily without being obscured by the magnet or the demonstrator's hand. The stream of moving negative ions is in fact deflected into a circular path, as theory predicts.

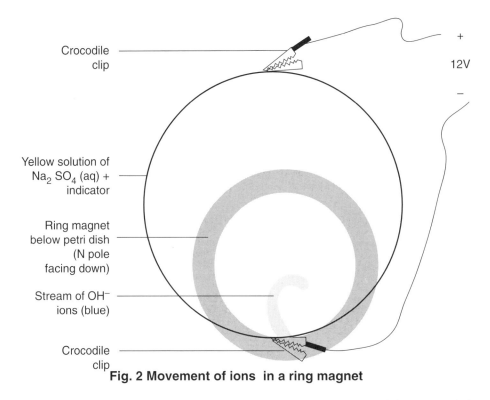

Fig. 2 Movement of ions in a ring magnet

A second possibility is to make a ring-shaped anode from a strip of aluminium foil placed around the inside of the rim of the petri dish. Stand the dish on the OHP above the ring magnet so that the dish and the magnet are concentric. Connect the aluminium foil anode to the positive terminal of the power pack with a lead and crocodile clip. Connect a second lead and crocodile clip to the negative terminal to act as the cathode and dip this crocodile clip into the solution in the centre of the petri dish (*Fig. 3*). The stream of indicator will travel outwards from the cathode in a spiral path in the direction predicted by the left hand motor rule.

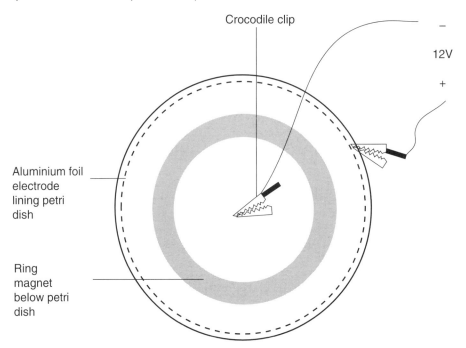

Fig. 3 Aluminium foil ring electrode

THE ROYAL
SOCIETY OF
CHEMISTRY

Visual tips

Adjust the amount of indicator so that the colour is seen easily when projected onto the screen. Use as little acid as possible to turn the indicator yellow. This will result in a darker blue colour when OH⁻(aq) ions are formed. Clean the OHP thoroughly and focus it carefully. Make sure any currents in the water caused by stirring have stopped before starting the demonstration. Method two is much better for visibility and should be used if a ring magnet is available.

Teaching tips

Students and, perhaps some teachers, will need to be reminded of the left hand motor rule (*Fig. 4*). Stress that the deflection of the ions is to the left or right *ie* at right angles to the magnetic field. The ions are not being attracted towards the pole of the magnet as a piece of iron would be, so the ions are not 'magnetic'.

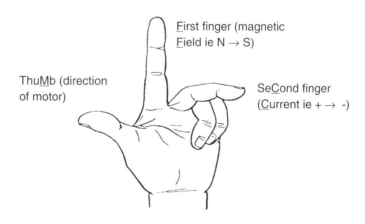

First finger (magnetic
Field ie N → S)

ThuMb (direction
of motor)

SeCond finger
(Current ie + → -)

Fig. 4 The left hand motor rule

Students may not be familiar with bromocresol purple and its colour changes. This indicator has been selected because it gives a good contrast on the OHP but others can be used if desired.

If appropriate, point out to students that this experiment is similar to the situation in a mass spectrometer except that there the ions are positive (and in the gas phase).

Students could be asked to predict the direction of deflection of the ions before it is demonstrated.

Some trials teachers reported that they used the demonstration simply to show the movement of ions during electrolysis and omitted the magnetic deflection.

Theory

A stream of moving anions is equivalent to a conventional (positively charged) current in the opposite direction. A charged particle moving in a uniform magnetic field is deflected in a circle as the force on it is always at right angles to its direction of motion.

At the cathode, H^+ (aq) ions are discharged:

$$2H_2O(l) + 2e^- \rightarrow H_2(g) + 2OH^-(g)$$

leaving behind OH⁻(aq) ions so the solution around the cathode becomes alkaline. The OH⁻(aq) ions are repelled from the cathode.

THE ROYAL
SOCIETY OF
CHEMISTRY

Extensions

Other indicators could be used *eg* phenolphthalein. Other electrolytes can be used
and the nature and shape of the electrodes can be changed. The voltage can be
varied as can the concentration of the solution. All of these extensions could be done
as student projects.

Safety

Wear eye protection.
   Take care not to spill water onto the OHP's electrics.
   The solution can be rinsed down the sink after use.
   It is the responsibility of teachers doing this demonstration to carry out an
appropriate risk assessment.

Acknowledgement

This demonstration is based on ideas from Colin Johnson at Techniquest, Cardiff, and
Brian Gray at the University of the Western Cape, South Africa.

THE ROYAL
SOCIETY OF
CHEMISTRY

44. Endothermic solid-solid reactions

Topic

Thermochemistry.

Timing

About 5 min.

Level

Pre-16 or post-16 if thermodynamic calculations are to be done.

Description

Stoichiometric quantities of solid barium hydroxide-8-water and solid ammonium chloride (or ammonium thiocyanate) are mixed in a beaker. Reaction takes place to produce a liquid, and the temperature drops to below -20 °C (-25 °C with ammonium thiocyanate).

Apparatus

▼ One 100 cm³ beaker.

▼ Watch glass.

▼ Thermometer reading to -30 °C, a thermocouple-type that can be connected to a large display or computer monitor is preferable.

▼ Access to a top pan balance.

▼ Access to a fume cupboard (optional).

Chemicals

The quantities given are for one demonstration.

▼ 32 g of **barium hydroxide-8-water** ($Ba(OH)_2.8H_2O$).

▼ 10 g of **ammonium chloride** (NH_4Cl) or 16 g of **ammonium thiocyanate** (NH_4SCN).

▼ A little **concentrated hydrochloric acid**.

▼ Universal indicator or litmus paper.

Method

Before the demonstration
Weigh out separately the barium hydroxide and the ammonium chloride (or ammonium thiocyanate). Avoid lumps as far as possible.

The demonstration
Stand the beaker on a watch glass containing a few drops of water. Note the room temperature. Mix the two solids in the beaker and stir with the thermometer. A liquid is formed – a white suspension with ammonium chloride and a colourless solution with ammonium thiocyanate. The presence of ammonia can be detected by smell, and confirmed by blowing fumes from the hydrochloric acid bottle across the beaker's mouth; and by using moist indicator paper. Work in a fume cupboard unless the room is well ventilated. Observe the drop in temperature, which is confirmed by the fact that the beaker freezes to the watch glass.

THE ROYAL
SOCIETY OF
CHEMISTRY

Visual tips

A large thermometer display helps. Pass the cold beaker around the class once the evolution of ammonia has stopped.

Teaching tips

Post-16 students could be asked to calculate the value of ΔH for the reaction from data book values for the ΔH_fs of the reactants and products. A value of $+164$ kJ mol^{-1} is obtained if the product is assumed to be $BaCl_2(s)$ and $+135$ kJ mol^{-1} if it is assumed to be $BaCl_2.2H_2O(s)$. Students should be able to predict that the reaction has a positive entropy change because a gas and a liquid are formed from two solids. They could also be asked to calculate the entropy change for the system and hence ΔS_{total} or ΔG for the reaction and thus confirm that the positive entropy change of the system outweighs the positive value of ΔH. A value of ΔS_{system} of $+591$ J mol^{-1}K^{-1} is obtained if the product is assumed to be $BaCl_2(s)$ and $+530$ J mol^{-1}K^{-1} if it is assumed to be $BaCl_2.2H_2O(s)$.

Theory

It is not possible to determine easily the exact barium compound or compounds produced in this reaction but the equation can be represented as

$$Ba(OH)_2.8H_2O(s) + 2NH_4Cl(s) \rightarrow 2NH_3(g) + 10H_2O(l) + BaCl_2(s)$$

or

$$Ba(OH)_2.8H_2O(s) + 2NH_4Cl(s) \rightarrow 2NH_3(g) + 8H_2O(l) + BaCl_2.2H_2O(s)$$

Extensions

Do other ammonium salts also react endothermically with barium hydroxide? Solid calcium hydroxide will react with solid ammonium chloride or ammonium thiocyanate to give off ammonia, but the reactions are much slower than those with barium hydroxide and are less endothermic. Water must be added to speed up the reactions to give a significant temperature drop.

Further details

The data required for the thermodynamic calculations are given below.

Compound	ΔH_f / kJ mol^{-1}	ΔS / J mol^{-1} K^{-1}
$Ba(OH)_2.8H_2O(s)$	-3345	427
$NH_4Cl(s)$	-314	95
$NH_3(g)$	-46	192
$H_2O(l)$	-286	70
$BaCl_2(s)$	-859	124
$BaCl_2.2H_2O(s)$	-1460	203

Safety

Wear eye protection.

The products can be disposed of by washing down the sink with large volumes of water.

It is the responsibility of teachers doing this demonstration to carry out an appropriate risk assessment.

45. A solid-solid reaction

Topic

Chemical reactions and reaction rates. Kinetic theory – the movement of particles in solids and liquids.

Timing

About 2 min.

Level

Lower secondary.

Description

Solid lead nitrate and solid potassium iodide are shaken together and yellow lead iodide is formed.

Apparatus

▼   One small screw-top jar.

Chemicals

The quantities given are for one demonstration.

▼   About 20 g of **lead(II) nitrate** (lead nitrate, $Pb(NO_3)_2$) and about 20 g of potassium iodide (KI).

Method

The demonstration

Weigh out equal masses of both compounds. These are then in approximately the stoichiometric ratio. Between 10 g and 20 g of each is suitable. Mix the solids in a screw topped jar and shake for several seconds. The yellow colour of lead iodide will be seen.

   Make a little more of the mixture and place it quickly into a beaker containing a little water. The reaction will take place much more rapidly.

Visual tips

The demonstration might have more impact if the jar is opaque and the yellow product can be poured out and shown to the unsuspecting audience. Have a white background available.

Teaching tips

Point out that for a reaction to occur, particles of the reactants must meet. This is much easier in solution (where particles are free to move) than in the solid state.

Theory

The reaction is:

$$Pb(NO_3)_2(s) + 2KI(s) \rightarrow 2KNO_3(s) + PbI_2(s)$$

All of these compounds are white except lead iodide, which is yellow.

Extensions

Lead ethanoate can be substituted for lead nitrate, but the reaction is much slower.
Are there any other examples of fairly rapid solid-solid reactions?

Safety

Wear eye protection.

The resulting mixture from the demonstration should be retained in a sealed container for professional disposal.

It is the responsibility of teachers doing this demonstration to carry out an appropriate risk assessment.

THE ROYAL
SOCIETY OF
CHEMISTRY

46. 'Magic' writing

Topic

This demonstration will probably be done for entertainment/general interest. As such it can be an excellent introduction or finale to a lecture demonstration programme if a suitable message is written. However, there is a lot of interesting chemistry going on – especially of transition metals.

Timing

A couple of minutes, but more if discussion is to take place and repeats are to be done.

Level

Any for general interest. Post-16 students should be able to appreciate the chemistry behind the reactions.

Description

Messages are written on filter paper with a variety of colourless, dilute, aqueous solutions. Spraying with other solutions produces coloured products and the messages show up in a variety of colours.

Apparatus

▼ Large sheets of white filter paper, chromatography paper or blotting paper.

▼ Small paintbrushes, the size used for painting models (wooden spills will do if these are not available).

▼ Spray bottles such as those used to spray house plants. These are available from garden centres and DIY shops. One is needed for each solution to be sprayed so three are required for the basic method described.

▼ Hair drier (optional).

Chemicals

The quantities given are sufficient for several demonstrations.

▼ 5 g of potassium hexacyanoferrate(II)-3-water (potassium ferrocyanide-3-water, $K_4Fe(CN)_6.3H_2O$).

▼ 5 g of **copper(II) sulphate-5-water** (hydrated copper sulphate, $CuSO_4.5H_2O$).

▼ 5 g of **ammonium thiocyanate** (NH_4SCN).

▼ 5 g of **iron(III) nitrate-9-water** (ferric nitrate-9-water, $Fe(NO_3)_3.9H_2O$).

▼ 5 g of **lead(II) nitrate** (lead nitrate, $Pb(NO_3)_2$).

▼ 5 g of potassium iodide, KI.

▼ 100 cm³ of approximately 2 mol dm⁻³ **ammonia solution** (bench ammonia, $NH_3(aq)$).

▼ A few cm³ of phenolphthalein solution.

▼ About 1 dm³ of deionised water.

Method

Before the demonstration

Make up the solutions as follows.

▼   Phenolphthalein: dissolve 0.1 g in 60 cm³ of ethanol and make up to 100 cm³ with deionised water. This is the usual bench solution.

▼   Ammonia: approximately 10 cm³ of 880 ammonia made up to 100 cm³ with deionised water.

▼   All the other solids: dissolve 5 g of solid in 100 cm³ of water.

Put the solutions of iron(III) nitrate, lead nitrate and ammonia into separate spray bottles. Adjust the nozzles of the bottles to give a fine mist and spray several times to ensure that the spray contains the solution and not water remaining from washing out the bottle.

Use paintbrushes or wooden spills to write suitable messages on the filter paper using solutions of potassium iodide, potassium hexacyanoferrate(II), ammonium thiocyanate, phenolphthalein and copper sulphate. Dry with a hair-drier. Work on a piece of clean newspaper to avoid picking up chemicals from the surface of the bench. All of the solutions will dry colourless with the exception of copper sulphate which will be a very pale blue – undetectable except to the most sharp-eyed audience. Pin the paper to the wall where it can be easily seen.

The demonstration

Spray the paper with the solution of lead nitrate. The message written in potassium iodide will show up as bright yellow lead iodide.

Now spray with the iron(III) nitrate. The message written with potassium hexacyanoferrate(II) will turn dark blue (Prussian blue) and the one written with ammonium thiocyanate will turn red-brown ($Fe(H_2O)_5SCN^{2+}$).

Now spray with the ammonia solution. The phenolphthalein will turn pink, the copper sulphate blue ($Cu(NH_3)_6^{2+}$) and the colour of the iron thiocyanate complex will disappear due to the formation of iron(III) hydroxide which is much less strongly coloured. Do not overspray the paper with the reagents as the colours will run.

Visual tips

A variety of suitable messages could be used such as 'Welcome' or 'The end' *etc* as appropriate. Alternatively, an equation could be written up so that the reactants are revealed by the first spray, the products with the second, and balancing coefficients with the third.

Teaching tips

Many of the reactions are worth discussing with a suitable audience. The reactions with ammonium thiocyanate and potassium hexacyanoferrate(II) are used as tests for iron(III).

Even fairly young students will know the reaction of phenolphthalein with alkali. Students could be encouraged to suggest that the pink writing could be erased with a dilute acid such as hydrochloric acid. This could be tried later.

THE ROYAL
SOCIETY OF
CHEMISTRY

Extensions

One interesting variation is to spray with a mixture of lead nitrate and iron(III) nitrate. This brings out the yellow, blue and brown colours simultaneously.

A large number of colour reactions can be used as the basis for 'magic writing' demonstrations. Do try them out first, especially when mixtures are involved, as unexpected reactions might embarrass the unwary.

Further details

Instead of spraying the solutions onto the filter paper, they could be painted on with large paintbrushes.

Safety

Wear eye protection.

Take care to avoid breathing in the fine sprays. Ensure that the room is well ventilated or do the demonstration outside or in a fume cupboard. Consider wearing a mask over the mouth.

It is the responsibility of teachers doing this demonstration to carry out an appropriate risk assessment.

THE ROYAL
SOCIETY OF
CHEMISTRY

47. The photochemical reactions of chlorine with hydrogen and with methane

Topic

Radical chain reactions, or unusual and spectacular exothermic reactions.

Timing

About 5 min.

Level

Any, as examples of interesting reactions. Post-16 as examples of chain reactions.

Description

Light from a slide projector is shone onto a small, corked plastic bottle containing a mixture of hydrogen and chlorine. This initiates a rapid photochemical reaction, the mixture explodes with a loud crack and the cork is fired into the air. Hydrogen chloride can be identified after the reaction.

Apparatus

▼ Slide projector with a bulb of rating 300 W or greater, or an electronic photographic flash gun.

▼ One 60 cm³ polyethylene bottle and cork (not a rubber bung) to fit. Suitable bottles can be obtained from equipment suppliers and many cosmetics products such as bath oil are sold in this type of bottle. The bottles appear to be almost opaque but they work well.

▼ Black insulating tape.

▼ Chlorine generator (500 cm³ conical flask with delivery tube or side arm and tap funnel), if a chlorine cylinder is not available.

▼ Two troughs or plastic washing up bowls for collecting the gases over water.

▼ Flexible delivery tubing for the gases.

▼ Stand with boss and clamp.

▼ Safety screen.

▼ Ear protection for the demonstrator.

▼ Access to a fume cupboard.

Chemicals

The quantities given are for one demonstration.

▼ **Hydrogen** cylinder with valve and regulator.

▼ **Chlorine** cylinder or 10 g of **potassium permanganate** (potassium manganate(VII), $KMnO_4$) and about 50 cm³ of **concentrated hydrochloric acid.**

▼ A bottle of **880 ammonia solution**.

▼ Blue litmus paper.

Method

Before the demonstration

Determine the volume of the plastic bottle by filling it with water and pouring the water into a measuring cylinder. Pour half this volume of water back into the bottle and mark the level with a permanent marker pen to give the half-way mark of the bottle. Wrap the bottle with black insulating tape leaving a 'window' about 2 cm x 1 cm centred on the half-way mark. Arrange the projector and the clamp so that the lens of the projector is about 5 cm from where the bottle will be clamped and the beam of the projector points directly at the window on the bottle. Place a safety screen between the bottle and the audience. Prepare the hydrogen cylinder and have ready a trough of water and a delivery tube. Prepare the chlorine cylinder or generator *(see Fig.1)* in the fume cupboard and have ready a trough of water and a

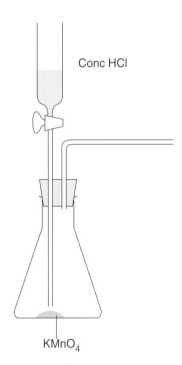

Conc HCl

KMnO$_4$

Fig. 1 The chlorine generator

delivery tube. Purge the tubes of both gas delivery systems so that they contain no air. The demonstration room should have subdued lighting to prevent full sunlight falling on the area where the demonstration will be done.

The demonstration

Fill a full bottle of water to the halfway mark with hydrogen by displacement of water. Displace the remaining water with chlorine. Cork the bottle. If there is any chance of bright sunlight impinging on the bottle, its window can be covered temporarily with a strip of black tape to eliminate any possibility of the reaction being initiated unexpectedly.

Clamp the bottle so that the cork points vertically and its window is facing the projector and about 5 cm from it (see Fig. 2). Ensure that the mouth of the bottle is not pointing at any light fittings. Switch the projector on. Almost immediately there will be a loud crack and the cork will be fired out of the bottle with some force and hit the ceiling. It is preferable to turn the projector on using the switch at the mains socket in order to be some way away from the bang. Demonstrators with sensitive ears will consider wearing ear protection and it is wise to insist that members of the audience put their fingers in their ears.

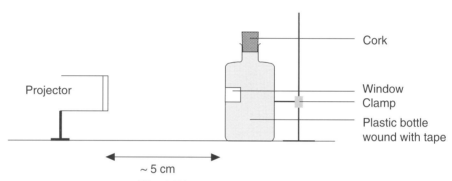

Fig. 2 Photochemistry of chlorine

The product, hydrogen chloride, can be identified by blowing fumes from the 880 ammonia solution bottle over the plastic bottle and by putting moist red litmus paper in its mouth.

The reaction can be initiated with the flash from an electronic photographic flash gun placed a few cm from the bottle's window. No camera is needed – the flash can be triggered by shorting its terminals. The flash from a compact automatic camera will not work. Alternatively the reaction can be started by holding a strip of burning magnesium a few cm from the window but this is more fiddly, less reliable, and tends to distract the audience from the main reaction.

Teaching tips

After the demonstration, get a member of the audience to place his or her hand in the projector beam to show that there is very little heat so that it must be light rather than heat that initiates the reaction. Students could be asked to calculate the wavelength of light needed to initiate the reaction and ΔH for the reaction from bond energies.

Theory

Both reactions are radical chain reactions in which the initiation step is the absorption of a photon by a chlorine molecule:

$$Cl_2(g) \xrightarrow{\ h\nu\ } 2Cl\bullet(g)$$

The Cl-Cl bond energy is 243 kJ mol^{-1}. This energy corresponds to a wavelength of 5×10^{-7} m – in the blue-green region of the visible spectrum.

THE ROYAL
SOCIETY OF
CHEMISTRY

In the chlorine-hydrogen reaction, this is followed by propagation steps:

$$Cl\bullet(g) + H_2(g) \rightarrow HCl(g) + H\bullet(g)$$

$$H\bullet(g) + Cl_2(g) \rightarrow HCl(g) + Cl\bullet(g)$$

Termination takes place via a variety of reactions such as:

$$2Cl\bullet(g) \rightarrow Cl_2(g)$$

$$H\bullet(g) + Cl\bullet(g) \rightarrow HC\,l(g)$$

and

$$2H\bullet(g) \rightarrow H_2(g)$$

which take place on the wall of the vessel to carry away excess energy.
Oxygen molecules can act as inhibitors via reactions such as:

$$H\bullet(g) + O_2(g) \rightarrow HO_2\bullet(g)$$

So the bottle must be carefully filled and corked to prevent oxygen from entering.

Since, in the overall reaction:

$$H_2(g) + Cl_2(g) \rightarrow 2HCl(g)$$

there is no increase in the number of moles of gas, the explosion must be due to the heat given out (93 kJ mol^{-1} of HCl) causing the gas mixture to expand.

Extensions

The reaction of chlorine with methane can be demonstrated in a similar way. The reaction starts less readily than with hydrogen and it is better to use a bottle with no tape. Otherwise the method is the same except that methane is used instead of hydrogen. A red flash will be seen in the bottle as the reaction takes place and the bottle becomes filled with a sooty deposit. The reaction is much less violent than that with hydrogen. Neither a flash gun nor burning magnesium ribbon will reliably initiate this reaction.

Further details

The steps in the photochemical reaction of chlorine and methane are given in a number of readily available sources. For example, *Revised nuffield advanced science: chemistry, students' book I*, p 273. London: Longmans, 1984.

One trials teacher generated the gases by the electrolysis of dilute hydrochloric acid.

Another stored the gases (separately) in gas syringes before filling the plastic bottle. The 'Gas bags ' (demonstration 19) could also be used.

Safety

Wear eye protection.

The hydrogen /chlorine reaction produces a loud bang and ear protection for the demonstrator is recommended. Students should be told to place their fingers in their ears.

THE ROYAL
SOCIETY OF
CHEMISTRY

There is some small risk of the bottle of mixed gases going off if exposed to sunlight, but experience suggests that this is minimal. It should be preventable if the window in the bottle is covered with tape. In any case it makes sense to make sure that the cork is pointed away from anyone while the bottle is being transferred. Should the mixture fail to explode, remove the cork and expel the gas mixture by inverting the bottle under a trough of water.

It is the responsibility of teachers doing this demonstration to carry out an appropriate risk assessment.

THE ROYAL
SOCIETY OF
CHEMISTRY

48. Dyeing – three colours from the same dye-bath

Topic

Dyes and dyeing.

Timing

15–30 min.

Level

Almost any for interest and entertainment. Post-16 if the structures of dyes and the types of bonding between dye and fabric are to be discussed.

Description

Samples of different fabrics are placed in a single dye bath containing three dyes. The materials emerge dyed different colours.

Apparatus

▼ Four 400 cm³ beakers.

▼ Four large watch glasses or petri dishes.

▼ Tongs or tweezers.

▼ Scissors.

▼ Access to a top pan balance.

▼ Bunsen burner, tripod, gauze and heat proof mat.

▼ String and paper clips, crocodile clips or clothes pegs to make a 'washing line'.

Chemicals

The quantities given are for one demonstration.

▼ Samples of the following fabrics in white: wool, silk, nylon, cotton, polyester, cellulose acetate ('triacetate'), polyester/cotton mix. About 100 cm² of each fabric or a few cm of thread will be sufficient. Nylon can be difficult to obtain and it may be necessary to try a second hand clothes shop.

▼ 0.05 g of each of the following dyes: acid blue 40, disperse yellow 7, direct red 23. These are available from Philip Harris.

▼ A little dilute **hydrochloric acid**.

Method

Before the demonstration
Cut four strips of each material (about 4 cm x 4 cm is suitable). Each fabric should be easily identifiable in some way for example by cutting different shapes. Weigh out two samples of 0.02 g of each of the red and yellow dyes and two samples of 0.03 g of the blue dye.

THE ROYAL
SOCIETY OF
CHEMISTRY

The demonstration

Dissolve 0.02 g of each of the red and yellow dyes and 0.03 g of the blue dye in 200 cm³ of water, add a few drops of dilute hydrochloric acid and heat to boiling. Place a sample of cotton, cellulose acetate and either wool, silk or nylon in the dye bath and simmer gently for about ten minutes. Remove the fabrics with tweezers or tongs and rinse under running water. Cotton will be dyed red, acetate yellow and wool, silk or nylon blue-green. (Some of the yellow direct dye will take to these materials as well as the blue acid dye). Try other materials as well if desired. Polyester will be dyed yellow and polyester/cotton will become orange.

Now examine the effect of the dyes individually. Make three dyebaths, the first containing 0.02 g of red dye in 200 cm³ of water, the second containing 0.02 g of the yellow dye in 200 cm³ of water and the third containing 0.03 g of the blue dye in 200 cm³ of water. Add a couple of drops of hydrochloric acid to each dye bath. Place a sample of each fabric in each dye bath and treat as before, *ie* simmer for ten minutes, remove the samples and rinse.

Typical results are shown in the table.

Dyes	Silk	Wool	Nylon	Cotton	Acetate	Polyester	Polycotton
Mix	olive-green	olive-green	olive-green	red	yellow	yellow	orange
Red	pale orange-red	pale orange-red	pale orange-red	red	almost white	pink	pink
Blue	blue	blue	blue	very pale blue	white	white	almost white
Yellow	orangey	orangey	orangey	pale yellow	bright yellow	bright yellow	bright yellow

Visual tips

Larger samples of material in larger volumes of dye can be used if the audience is large. A 'washing line' is useful, on which to hang samples to dry with clothes pegs, crocodile clips or paper clips.

Teaching tips

After seeing the action of the dyes on, say, wool, students could be asked to predict the effect on silk and nylon – which are also polyamides. After seeing the effect of the dyes on polyester and cotton separately, students could be asked to predict their effect on the mixed fabric. Point out that this experiment can help explain some odd effects in washing machine accidents where labels and trim may emerge a different colour to the rest of the garment.

Theory

Different dyes bond to fabrics in different ways.

Acid dyes contain acidic $-CO_2H$ and $-SO_3H$ groups which bond to the basic $-NH$ groups in the amide linkages of wool, silk and nylon.

Direct dyes bond by hydrogen bonding and take well to cellulose-based fibres such as cotton, viscose and rayon which have many $-OH$ groups.

Disperse dyes are not water-soluble. They exist in the dye-bath as a fine suspension (hence the name), and are absorbed as a solid solution by hydrophobic fabrics such as polyesters.

THE ROYAL
SOCIETY OF
CHEMISTRY

The structures of the dyes used in this experiment are:

Direct red 23

Disperse yellow 7

Acid blue 40

Extensions

Mixed dye-baths that produce different colours to the ones suggested here can be devised using the principles governing what dyes colour each of the fabrics described above.

Do mordants such as salt or alum have any effect? Does pH have any effect? Does the time in the dye bath or the temperature of the dye bath have any effect? How fast are the dyes to a variety of treatments?

Further details

There are, of course, other types of dye. A good account of which types of dyes dye which fabrics is given in *The essential chemical industry,* p 42. University of York: The Chemical Industry Education Centre, 1989.

A number of alternative methods for this type of experiment is given in an ICI booklet – *School link the 1+1 = 3 challenge* although this is no longer available from Zeneca (formerly ICI). A kit of dyes and fabrics for a similar demonstration is available from Kemtex Services Ltd, Tameside Business Centre, Windmill Lane, Denton, Manchester M34 3QS. Tel: 0161 320 6505.

Safety

Wear eye protection.

These dyes may irritate some people's skins and their dusts may irritate the respiratory tract. Plastic gloves are recommended. Open the bottles in the fume cupboard.

It is the responsibility of teachers doing this demonstration to carry out an appropriate risk assessment.

THE ROYAL
SOCIETY OF
CHEMISTRY

49. The reaction of sodium with chlorine

Topic

Periodic Table. Group I metals and their reactions. Group VII elements and their reactions.

Timing

2 - 3 min.

Level

Pre-16.

Description

A little sodium is placed in a gas jar of chlorine. The reaction is initiated by placing a drop of water on the metal. A vigorous reaction follows and the white product (sodium chloride) is easily visible.

Apparatus

▼   If a chlorine cylinder is not available, a chlorine generator will be required (500 cm³ conical flask with side arm or delivery tube fitted with a one-holed bung and a tap funnel).

▼   Gas jar with lid or 500 cm³ conical flask with bung to fit.

▼   Small piece of ceramic material, such as a piece of an evaporating basin or heat-proof mat, to fit inside the gas jar or flask.

▼   Dropping pipette or wash bottle.

▼   Knife to cut the sodium.

▼   Filter paper or paper towels to wipe the oil from the sodium.

▼   Access to a fume cupboard.

Chemicals

The quantities given are for one demonstration.

▼   **Chlorine** cylinder or 10 g of **potassium permanganate** (potassium manganate(VII), $KMnO_4$) and about 50 cm³ of **concentrated hydrochloric acid**.

▼   A piece of **sodium** about half the size of a pea.

▼   A little **silver nitrate** solution.

▼   A few cm³ of a hydrocarbon solvent such as **hexane** to clean the oil from the sodium.

▼   A little vaseline for the gas jar lid.

▼   Tweezers.

Method

Before the demonstration
Set up the chlorine generator or cylinder in the fume cupboard.

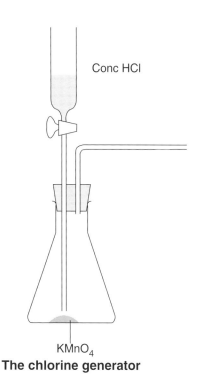

Conc HCl

$KMnO_4$

The chlorine generator

The demonstration

Place the piece of ceramic material in the base of the gas jar or flask. This protects the glass from cracking with the heat of the reaction. Fill the jar or flask with chlorine by upward displacement of air and put on the lid or bung. The contents of the flask will be green. Cut a piece of sodium about half the size of a pea, clean the oil from it using a little hexane (or other suitable solvent) and dry with a filter paper. Using tweezers, place the sodium on the ceramic material and place one drop of water on the sodium. This will react exothermically with the sodium and the heat generated will be sufficient to start the reaction between the sodium and the chlorine. The sodium will burn brightly in the chlorine with an orange flame, clouds of white sodium chloride will be seen and the green colour of the chlorine will disappear. When the reaction is over, carefully add a few cm^3 of water (as all the sodium may not have reacted). Add a little silver nitrate solution and observe the white precipitate showing the presence of a chloride.

Visual tips

A white background helps to make the colour of the chlorine more easily visible.

Teaching tips

Explain that the water is used only to provide heat to start the reaction. It will be helpful if the audience is already familiar with the sodium-water reaction.

Extensions

This method works with potassium (which may ignite spontaneously on being placed in the chlorine) but not with lithium.

Further details

This method of demonstrating the reaction avoids the formation of brown iron(III) chloride which occurs if an iron deflagrating spoon is used in the usual way. An alternative is to improvise a glass deflagrating spoon by cutting off the bottom 2 mm

THE ROYAL
SOCIETY OF
CHEMISTRY

of a test-tube and holding this in tweezers or tongs. It may be necessary to break the test-tube and use the bottom fragment if suitable glass cutting equipment is not available.

The article *The exploding metals* by G. D. John, *Sch. Sci. Rev.*, 1980, **62** (219), 279, discusses many aspects of the reactions of the alkali metals.

Safety

Wear eye protection.

It is the responsibility of teachers doing this demonstration to carry out an appropriate risk assessment.

THE ROYAL
SOCIETY OF
CHEMISTRY

50. Unsaturated compounds in foods

Topic

Food chemistry. Organic chemistry – addition reactions of alkenes.

Timing

About 5 min.

Level

Post-16, but the topic of unsaturation in fats is of general interest.

Description

Various fatty food products are placed in flasks containing bromine vapour. The colour of the bromine disappears indicating that the foods contain carbon carbon double bonds.

Apparatus

▼ Several 1 dm³ flasks with rubber bungs to fit. One more flask is required than the number of foods to be tried.

▼ A dropping pipette.

▼ Bunsen burner, tripod and gauze.

▼ Tongs.

▼ One 100 cm³ beaker.

▼ Access to a fume cupboard.

Chemicals

The quantities given are for one demonstration.

▼ 1 cm³ of liquid **bromine**.

▼ One rasher of bacon.

▼ A few cm³ of cooking oil.

▼ A small knob of margarine.

Method

Before the demonstration

Working in a fume cupboard, open an ampoule of bromine and transfer the bromine to a small bottle. Using a dropping pipette place two or three drops of bromine into each of four 1 dm³ flasks and stopper them. Shake the flasks until all the bromine has vaporised and there is an easily visible brown coloration in each flask.

The demonstration

For bacon:

Hold about a quarter of a rasher of bacon in tongs and heat it in a Bunsen flame until it is sizzling. Put the bacon in one of the bromine-filled flasks, re-stopper the flask and shake. After several seconds the brown colour of the bromine will have disappeared.

THE ROYAL
SOCIETY OF
CHEMISTRY

For cooking oil:
Pour a few cm³ of the oil into one of the bromine-filled flasks, re-stopper the flask and shake. After several seconds the brown colour of the bromine will have disappeared.

For margarine:
Place a small knob of margarine in a small beaker and heat it over a Bunsen burner until it has melted. Pour the melted margarine into a flask and shake. After several seconds the brown colour of the bromine will have disappeared.

Visual tips

A white background will help. Use the extra bromine-filled flask as a control to make the disappearance of the colour more striking.

Teaching tips

Have available the labels of the products, where possible, to pass round the class. Point out and explain the terms unsaturates, polyunsaturates and monounsaturates.

Theory

Bromine adds on across the carbon-carbon double bonds of unsaturated molecules:

$$\begin{array}{ccc} & & Br\quad Br \\ & & |\quad\quad| \\ \diagdown\diagup & & -C-C- \\ C=C \quad +Br_2 \;\rightarrow & & |\quad\quad| \\ \diagup\diagdown & & H\quad H \\ H\quad\quad H & & \end{array}$$

The unsaturated aldehyde propenal, $CH_2=CHCHO$, is reported to be one of a number of chemicals responsible for the odour of frying bacon, but unsaturated fats will be present as well.

Extensions

Try other food products eg lard, butter etc. Solid products could be liquefied by dissolving them in a suitable solvent such as hexane rather than melting them. In this case some bromine will dissolve in the solvent before reacting. It will be necessary to show that the solvent does not decolorise the bromine itself. It is wise to check this beforehand because sometimes solvents contain unsaturated impurities.

Further details

Some supermarkets produce leaflets on foods which might be relevant.

Safety

Wear eye protection.
    Bromine is toxic by inhalation, has an irritant vapour and causes severe burns. Because of its high density and high vapour pressure it can be tricky to transfer with a dropping pipette. Work with bromine in a fume cupboard and wear latex gloves. Have available a 1 mol dm⁻³ solution of sodium thiosulphate to deal with any spills.
    Dispose carefully of any food that remains after the demonstration.
    It is the responsibility of teachers doing this demonstration to carry out an appropriate risk assessment.

THE ROYAL
SOCIETY OF
CHEMISTRY

51. Making silicon and silanes from sand

Topic

The reactions of group IV elements. Comparison of hydrocarbons with silanes, with regard to thermodynamic and kinetic stability.

Timing

5 – 10 min.

Level

Pre-16 or post-16 depending on the level of discussion.

Description

Magnesium and sand are heated together and silicon is produced by an exothermic reaction. The product is placed in acid to remove magnesium oxide and unreacted magnesium. Small amounts of silanes are produced by the reaction of magnesium silicide (a side product) with the acid. These react spontaneously with air to give spectacular but harmless small explosions.

Apparatus

▼   One pyrex test-tube, approximately 150 mm x 17 mm.

▼   Clamp and stand.

▼   Bunsen burner.

▼   One 250 cm³ beaker.

▼   One 250 cm³ conical flask.

▼   Filter funnel and filter paper.

▼   Access to oven.

▼   Desiccator.

▼   Access to top pan balance.

▼   Safety screen.

Chemicals

The quantities given are for one demonstration.

▼   1 g of dry **magnesium powder**.

▼   1 g of dry silver sand.

▼   About 50 cm³ of approximately 2 mol dm⁻³ **hydrochloric acid**.

Method

Before the demonstration

It is important that the reactants are dry. Dry the magnesium powder and the sand for a few hours in an oven at about 100 °C. Store them in the desiccator until ready to use them. Ensure that the test-tube is dry.

The demonstration

Weigh 1 g of silver sand and 1 g of magnesium powder and mix them thoroughly. This mixture has a small excess of magnesium over the stoichiometric masses (1 g of sand to 0.8 g of magnesium) because some magnesium will inevitably react with air. Spread the mixture along the bottom of a test-tube that is clamped almost horizontally. Place a safety screen between the tube and the audience if the spectators are close.

Heat one end of the mixture with a roaring Bunsen flame, holding the burner by hand. After a few seconds the mixture will start to glow. This glow can be 'chased' along the tube with the flame until all the mixture has reacted. The tube will blacken and partly melt. If the two powders are not dry, some magnesium will react with the steam and the resulting hydrogen will pop. This can be disconcerting if it is not expected.

When the reaction is complete, allow the mixture to cool (about five minutes) and with the aid of a spatula pour the products into about 50 cm³ of 2 mol dm⁻³ hydrochloric acid to dissolve away unreacted magnesium and magnesium oxide. The solid will contain silicon, magnesium oxide (the main products), magnesium silicide formed from the reaction of excess magnesium with silicon, unreacted magnesium and possibly a little unreacted sand. The mixture will fizz as excess magnesium reacts with the acid. There will also be pops accompanied by small yellow flames. These are caused by silanes that are formed from the reaction of magnesium silicide with acid. Silanes inflame spontaneously in air. Magnesium oxide will dissolve in the acid.

After a few minutes the pops will cease and grey silicon powder, possibly with a little unreacted sand, will be left on the bottom of the beaker. Pour off the acid, wash the solid a few times with water and filter off the silicon. It can be passed around the class to show its slightly metallic silver-grey colour. If desired show that it does not react with alkalis (or acids).

Visual tips

Make sure the safety screen is clean if one is used.

Teaching tips

There are many interesting contrasts to be drawn between silicon compounds and their carbon analogues. Silicon dioxide is a solid with a giant structure, while carbon dioxide is molecular. Silanes react spontaneously with air at room temperature while alkanes are stable. These differences can be explained by considering the relevant bond energies and availability of d-orbitals in silicon but not in carbon.

Bond energies in kJ mol⁻¹: Si=O 638; Si–O 466; C–O 336; C=O 805; Si–H 318; C–H 413.

Theory

The reactions are:

$$SiO_2(s) + 2Mg(s) \rightarrow 2MgO(s) + Si(s)$$

$$2Mg(s) + Si(s) \rightarrow Mg_2Si(s)$$

$$MgO(s) + 2HCl(aq) \rightarrow MgCl_2(aq) + H_2O(l)$$

$$Mg_2Si(s) + 4HCl(aq) \rightarrow 2MgCl_2(aq) + SiH_4(g)$$
(Higher silanes such as Si_2H_6 may also be produced.)

$$SiH_4(g) + 2O_2(g) \rightarrow SiO_2(s) + 2H_2O(l)$$

THE ROYAL
SOCIETY OF
CHEMISTRY

Further details

One teacher reported that the 'pops' continued while the silicon dried on the filter paper.

Silicon is extracted from sand industrially by reduction with carbon.

Safety

Wear eye protection.

Use a safety screen between the apparatus and the audience.

Magnesium powder burns vigorously in air. The dust from magnesium powder may be hazardous. Ensure that the mixed powders are absolutely dry before the reaction.

It is the responsibility of teachers doing this demonstration to carry out an appropriate risk assessment.

52. Red, white and blue

Topic

Chemical reactions: acid-base, precipitation, complex formation. This demonstration will be of most use for general interest on open days *etc.*

Timing

2-3 min.

Level

Any, for general interest. Post-16 for a full understanding of the reactions.

Description

A solution of ammonia is poured into each of three beakers which contain (unknown to the audience) a little phenolphthalein, a little lead nitrate solution and a little copper sulphate solution respectively. The beakers' contents turn red, milky white and deep blue respectively. Pouring the contents of the beakers into acid reverses the changes to give a colourless solution.

Apparatus

▼   Three 250 cm³ beakers.

▼   Two identical flasks of about 500 cm³ capacity.

▼   Three teat pipettes.

Chemicals

The quantities given are for one demonstration.

▼   About 250 cm³ of 1 mol dm⁻³ **ammonia solution**. (To make 1 dm³ of 1 mol dm⁻³ ammonia solution, make 57 cm³ of 880 ammonia up to 1 dm³ with deionised water.)

▼   About 250 cm³ of 2 mol dm⁻³ **nitric acid**. (To make 1 dm³ of 2 mol dm⁻³ nitric acid, make 125 cm³ of **concentrated (70 %) nitric acid** up to 1 dm³ with deionised water.)

▼   About 1 cm³ of approximately 0.5 mol dm⁻³ copper sulphate. (Add about 2.5 g of **copper sulphate-5-water** to 10 cm³ of deionised water.)

▼   About 1 cm³ of saturated lead(II) nitrate solution. (Add 14 g of **lead nitrate** to 10 cm³ of deionised water to make a saturated solution.)

▼   About 1 cm³ of phenolphthalein solution. (This is made by dissolving 1 g of phenolphthalein solution in 600 cm³ of **ethanol** and making up to 1 dm³ with deionised water.)

Ensure that all of the solutions are made up using deionised water otherwise the lead-containing solutions will be cloudy due to the formation of lead chloride from chloride ions in tap water.

Method

Before the demonstration

Line up the three 250 cm³ beakers on the bench. Place about 1 cm³ of phenolphthalein solution in the first, place about 1 cm³ of saturated lead nitrate solution in the second and place about 1 cm³ of saturated copper sulphate solution in

THE ROYAL
SOCIETY OF
CHEMISTRY

the third. The volumes are not critical – a single 'squirt' from a teat pipette will be accurate enough. The audience should not know about these additions. Only the most sharp-eyed observers will notice even the copper sulphate. Place 250 cm³ of ammonia solution in one 500 cm³ flask and about 125 cm³ of the nitric acid in the other, which should be kept out of sight of the audience. Mark the ammonia flask at approximately the 125 cm³ level.

The demonstration

Pour about 40 cm³ of ammonia solution in turn into each of the three beakers on the bench. Aim to leave the flask full to the mark at 125 cm³. The phenolphthalein will turn red, the lead nitrate will form a milky white precipitate of lead(II) hydroxide and the copper sulphate will form the deep blue tetraamminecopper(II) ion. Now use some sleight of hand to switch the ammonia-containing flask with that containing the nitric acid. The levels of liquid in both flasks will now be about the same. Pour the contents of the three beakers in turn into the nitric acid flask and the colours will disappear, leaving a clear, colourless solution. (In fact it may be a very pale blue due to the copper ions and there may be a few specks of undissolved lead hydroxide, but the audience is unlikely to notice this.)

Visual tips

Scale the volumes up if the audience is some way away. Stand the phenolphthalein and copper sulphate flasks on white filter paper and the lead nitrate one on black paper for maximum impact.

Teaching tips

Go over the reactions with a suitable audience. Ask them to predict the contents of the second flask (after explaining the sleight of hand). Ask them to suggest ways of producing other colours.

Theory

The reactions are:

$$Pb(NO_3)_2(aq) + 2NH_3(aq) + 2H_2O(l) \rightarrow Pb(OH)_2(s) + 2NH_4NO_3(aq)$$

$$Cu^{2+}(aq) + 4NH_3(aq) \rightarrow Cu(NH_3)_4^{2+}(aq).$$

These are reversed in acid:

$$Pb(OH)_2(s) + 2HNO_3(aq) \rightarrow Pb(NO_3)_2(aq) + 2H_2O(l)$$

$$Cu(NH_3)_4^{2+}(aq) + 4H^+(aq) \rightarrow Cu^{2+}(aq) + 4NH_4^+(aq)$$

Further details

A version of this demonstration in which the evolution of nitrogen dioxide generated by the reaction of copper and concentrated nitric acid drives the solutions from one flask to another has been reported by T. C. Swinfen and D. J. Hearn, *Sch. Sci. Rev.,* 1989, **71** (255), 94.

Safety

Wear eye protection.

It is the responsibility of teachers doing this demonstration to carry out an appropriate risk assessment.

53. The reduction of copper oxide

Topic

Formula determination, extraction of metals.

Timing

About 10 min.

Level

Pre-16.

Description

Copper(II) oxide can be reduced by hydrogen and its formula determined. Natural gas (mainly methane) can also be used as a reducing agent, but the reaction is much slower. The reduction with methane can be speeded up by either bubbling the methane through ethanol or by placing a piece of firelighter in the gas stream in the reduction tube.

Apparatus

For the basic method

▼   One reduction tube, *ie* a Pyrex boiling tube with a small hole blown about 1 cm from the closed end (*Fig. 1*).

▼   A one-hole rubber bung to fit the reduction tube fitted with a short length of glass tubing.

▼   Rubber tubing to connect the reduction tube to the hydrogen cylinder or gas tap.

▼   Stand, boss and clamp.

▼   Bunsen burner.

▼   Access to a top pan balance that weighs to 0.01 g.

▼   Safety screen.

▼   Circuit board, batteries, bulb, ammeter and leads to test the electrical conductivity of the product (optional).

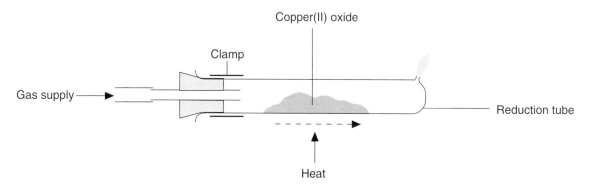

Fig. 1 Apparatus for copper(II) oxide reduction

THE ROYAL
SOCIETY OF
CHEMISTRY

For the bubbling through ethanol method

▼ One 250 cm³ conical flask fitted with a two-holed bung with one long and one short glass tube (*Fig. 2*).

▼ One 1 dm³ beaker.

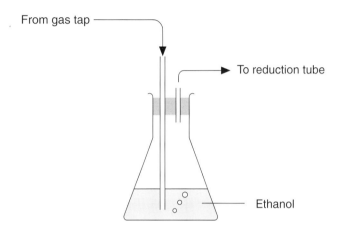

Fig. 2 Bubbling through ethanol apparatus

Chemicals

The quantities given are for one demonstration.

For the basic method

▼ About 3 g of **copper(II) oxide** (black copper oxide, CuO) (analytical grade is needed for good quantitative results – lesser grades of copper(II) oxide often contain significant amounts of copper).

▼ About 3 g of copper(I) oxide (red copper oxide, Cu_2O) (optional).

▼ **Hydrogen** cylinder with valve gear and regulator.

For the bubbling through ethanol method

▼ About 100 cm³ of **ethanol** is also needed.

For the firelighter method

▼ A small piece (about 1 cm x 1 cm x 2 cm) of white firelighter is also needed. This may be bought from Tesco.

Method

Before the demonstration

Dry the copper oxide in an oven at about 100 °C and store it in a desiccator until the demonstration.

The demonstration

a) With hydrogen

Weigh the reduction tube empty. Place about 3 g of copper(II) oxide along the base of the tube so that it is spread out over a length of about 4 cm centred on the middle of the tube. This is to ensure that it will not be necessary to heat too close to the rubber bung and so that there is no tendency for the powder to be blown out of the hole when the hydrogen is turned on. Reweigh and note the mass of the tube plus copper(II) oxide. Clamp the reduction tube at its open end (a clamp without cork or rubber padding is preferable as the tube may get hot). Place a safety screen between the tube and the audience. Connect the bung and glass tube to the hydrogen cylinder with rubber tubing, turn on the gas and adjust to get a gentle flow which can just be felt on the cheek. Place the bung in the mouth of the reduction tube. Check that the hydrogen is coming out of the hole in the reduction tube by collecting the gas in a micro test-tube and seeing if it burns quietly rather than 'pops' when placed in a Bunsen flame. Leave for a further 30 seconds to flush out the air, and light the gas emerging from the hole. Adjust the gas flow to give a flame about 3 cm high.

Light the Bunsen burner and with the tip of a roaring flame heat the oxide at one end of the pile. After a few seconds the powder will glow and start to turn pink. Chase this glow along the tube for about 30 seconds until the whole of the black oxide has turned to pink copper. Continue to heat the tube for at least another minute to ensure that all the oxide has reacted. Take care that the Bunsen flame does not extinguish the hydrogen flame and be prepared to re-light this flame if it goes out. Remove the Bunsen burner and allow the reduction tube to cool with the hydrogen still passing over the copper and the excess gas still burning. This prevents air coming into contact with the hot copper and converting it back to oxide. When the tube is cool enough to handle, turn off the hydrogen at the cylinder, remove the bung and weigh the tube and contents.

NB Wait until the flame has gone out before removing the bung. Otherwise, as air is drawn into the reduction tube, there will be a loud pop which can be disconcerting.

b) With natural gas
The method is as above but using gas from the gas tap (about 95 % methane) instead of hydrogen. The reduction is much slower – it will take about 20 minutes of strong heating – and no glow is seen. It is difficult to be certain when reduction is complete and teachers might consider reducing to constant weight, but the time taken for several cooling, weighing and re-heating cycles seems excessive for a demonstration. It is important that the whole of the oxide is heated strongly.

c) With natural gas and ethanol
As for (b) above, but arrange for the natural gas to bubble through ethanol in a conical flask before it reaches the reduction tube (*Fig. 2*). Standing the flask in warm water (about 65 °C) to increase the rate of evaporation of the ethanol speeds up the reduction.The reduction takes between five and ten minutes and a faint glow is seen as the oxide is reduced.

d) With natural gas and firelighter

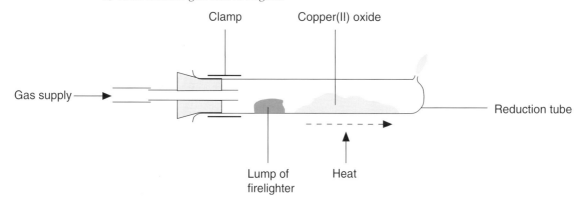

Fig. 3 Using a firelighter

THE ROYAL
SOCIETY OF
CHEMISTRY

As for *(b)* above but place a small piece (about 1 cm x 1 cm x 2 cm) of white firelighter in the reduction tube close to the gas inlet *(see Fig. 3)*. Do this after weighing the tube plus oxide. Heat the oxide strongly and flick the Bunsen flame occasionally onto the firelighter for a few seconds at a time. The firelighter contains a paraffin-like hydrocarbon which is vaporised by the heat. The firelighter blackens, the flame burning the excess gas goes smoky and the reduction tube may blacken slightly. The reduction takes about ten minutes. After cooling the tube remove the remains of the firelighter with a spatula before weighing.

Visual tips

Once the gas has been lit it should be possible to remove the safety screen for better visibility.

Teaching tips

During the reduction, many interesting observations can be pointed out to the audience:

▼ Before the tube gets too hot, droplets of water produced by the reaction can be seen near the end of the reduction tube.

▼ The excess hydrogen (or natural gas) flame will burn green – the characteristic colour of copper.

▼ During the reduction, the height of the flame will lessen as some hydrogen is being used up and replaced by water.

Pass the copper that is produced around the class. Demonstrate that it conducts electricity by using a circuit board and ammeter.

Formulae can be determined from the combining masses in the usual way or by using the graphical method described in *Revised Nuffield chemistry teachers' guide II p71*, London: Longman, 1978.

Some teachers may wish simply to demonstrate the removal of oxygen from an oxide rather than to use this procedure to determine formulae.

Theory

The reactions are:

$$CuO(s) + H_2(g) \rightarrow Cu(s) + H_2O(l)$$

$$4CuO(s) + CH_4(g) \rightarrow 4Cu(s) + 2H_2O(l) + CO_2(g)$$

$$6CuO(s) + C_2H_5OH(l) \rightarrow 6Cu(s) + 3H_2O(l) + 2CO_2(g)$$

Extensions

The formula of red copper(I) oxide can be determined by reduction with hydrogen in exactly the same way as for copper(II) oxide. The colour change is less clearcut, but the powder glows showing that reduction is taking place.

Oxides of lead can be reduced by this method using hydrogen, as can other metal oxides.

Safety

Wear eye protection.

Use a safety screen between the apparatus and the audience.

It is the responsibility of teachers doing this demonstration to carry out an appropriate risk assessment.

THE ROYAL
SOCIETY OF
CHEMISTRY

54. The ammonium dichromate 'volcano'

Topic

Exothermic reactions, general interest. It could be used to liven up an Earth science lesson but the resemblance to a volcano is coincidental.

Timing

About 2 min.

Level

Any for general interest. Post-16 if it is to be used along with thermodynamic calculations.

Description

A small conical heap of orange ammonium dichromate is ignited. It sparks and produces a large volume of green chromium(III) oxide as well as steam, resembling a volcano.

Apparatus

▼   Bunsen burner, heat-proof mat.

▼   Metal tray such as a large tea tray.

▼   Watch glass.

▼   Bell jar (optional).

▼   One 1 dm³ flask (optional).

▼   One 250 cm³ flask (optional).

Chemicals

The quantities given are for one demonstration.

▼   10 g of **ammonium dichromate(VI)** (ammonium dichromate, $(NH_4)_2Cr_2O_7$).

▼   Wooden spill.

▼   A little **ethanol**.

▼   One piece of blue cobalt chloride paper.

▼   A little glass wool or mineral wool (optional).

▼   A few grams of silica gel granules (optional).

▼   Access to a fume cupboard (optional).

Method

The demonstration

Working in a fume cupboard, place a conical pile of about 10 g of ammonium dichromate on a heat-proof mat standing on a tray to collect the chromium oxide that shoots into the air. Soak about a 3 cm length of wooden spill in ethanol and stick this into the top of the pile so that about 2 cm protrudes to act as a wick. Light the wick. As the wick burns down into the ammonium dichromate, the compound begins to spark and decompose leaving behind a cone of green chromium(III) oxide that has a

THE ROYAL
SOCIETY OF
CHEMISTRY

considerably larger volume than the original compound. Some of this oxide shoots into the air. The 'volcano' burns for between 30 seconds and one minute. Hold a watch glass above the 'volcano'; this becomes steamed up with water from the decomposition. Confirm that this is water with blue cobalt chloride paper.

The reaction may also be started by pointing a roaring Bunsen flame at the top of the pile of ammonium dichromate.

Visual tips

A portable fume cupboard gives all-round vision. If it is desirable to do the demonstration without a fume cupboard, place a large bell jar over the reaction. However, this soon steams up. Place matchsticks or something similar under the rim of the bell jar to allow the nitrogen produced in the reaction to escape.

Teaching tips

If appropriate, some students could be asked to predict the products given the formula of ammonium dichromate.

The demonstration could be used to enliven a lesson on thermodynamics (post-16) in which case students could be asked to calculate ΔH, ΔS and hence ΔG for the reaction. The values they should obtain are:

$\Delta H = -478 \text{ kJ mol}^{-1}$
$\Delta S = +217 \text{ J mol}^{-1} \text{ K}^{-1}$
$\Delta G = -543 \text{ kJ mol}^{-1}$

They should be able to predict qualitatively that there is an entropy increase. The data required are given in the table.

	ΔH_f / kJ mol^{-1}	ΔS / J mol^{-1} K^{-1}
$(NH_4)_2Cr_2O_7(s)$	−1806	336 (estimated)
$Cr_2O_3(s)$	−1140	81
$N_2(g)$	0	192
$H_2O(l)$	−286	70

Post-16 students could also be asked to balance the equation using oxidation numbers.

Extensions

An alternative way of doing the experiment without a fume cupboard is as follows. Place about 3 g of ammonium dichromate in a 1 dm³ conical flask. Place a loose plug of glass wool or mineral wool in the mouth of the flask to prevent loss of chromium(III) oxide. Start the reaction by heating the flask on a Bunsen burner with the tip of a roaring flame pointing at the pile of ammonium dichromate. Once the reaction has started, place the flask on a heat-proof mat in view of the audience. The flask will steam up somewhat and a little steam will escape. To confirm that this is a decomposition reaction rather than a combustion reaction, flush the flask with nitrogen from a cylinder and repeat the reaction. It will be unaffected.

It is possible to modify this method to suggest that a gas is formed. Replace the glass wool or mineral wool plug with a loose sandwich of glass wool and silica gel granules to absorb any steam (see figure). Weigh the flask before and after the reaction. A weight loss will be observed suggesting loss of gas (although it is difficult to ensure that no steam escapes).

Calculation shows that 2.5 g of ammonium dichromate should produce about 240 cm³ of nitrogen. React 2.5 g of ammonium dichromate in a 250 cm³ conical flask with a loose glass wool or mineral wool plug. Immediately the reaction has finished, place a lighted taper in the flask. The nitrogen will cause it to go out. Compare with a lighted taper in an air filled flask of the same size. It will burn for several seconds.

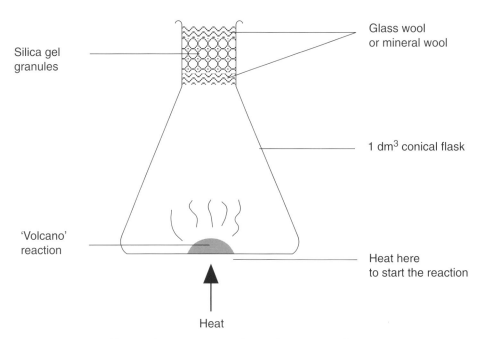

Alternative apparatus for volcano reaction

Theory

The reaction that occurs is:

$$(NH_4)_2Cr_2O_7(s) \rightarrow Cr_2O_3(s) + N_2(g) + 4H_2O(l)$$

Further details

The value for the standard molar entropy of ammonium dichromate has been estimated by Latimer's rules, which state that the entropy of each atom of each element in a compound in $JK^{-1} mol^{-1}$ is given by:

$$\Delta S^\circ = (1.5R \ln A_r) - 3.92$$

(where A_r is the relative atomic mass of the element and R the gas constant) and that the entropy of a compound is the sum of the entropies of all the elements in the compound.

For further details see W. M. Latimer, *J. Am. Chem. Soc.*, 1921, **43**, 818.

This rule may be found useful for estimating the entropies of other compounds which are not readily available in the literature.

Safety

Wear eye protection.

Dispose of the residue in a sealed plastic bag placed in the dustbin.

It is the responsibility of teachers doing this demonstration to carry out an appropriate risk assessment.

THE ROYAL
SOCIETY OF
CHEMISTRY

55. Sulphuric acid as a dehydrating agent

Topic

Properties of sulphuric acid.

Timing

Less than 5 min.

Level

Lower secondary.

Description

Concentrated sulphuric acid is added to sucrose in a beaker. The sucrose is dehydrated leaving a spongy mass of carbon which rises from the beaker like a 'serpent'.

Apparatus

▼ One 100 cm³ beaker.

▼ One large watch glass.

▼ One measuring cylinder to measure 20 cm³.

▼ One dropping pipette.

▼ Access to fume cupboard.

Chemicals

The quantities given are for one demonstration.

▼ 50 g of sucrose (table sugar).

▼ 20 cm³ of **concentrated sulphuric acid**.

▼ About 3 g of **copper sulphate-5-water (CuSO$_4$.5H$_2$O)**.

▼ About 50 g of glucose (optional).

▼ One piece of blue cobalt chloride paper.

▼ One piece of **potassium dichromate** paper (dip a strip of filter paper in a saturated solution of potassium dichromate).

Method

The demonstration

Weigh about 50 g of sucrose (ordinary table sugar) into a 100 cm³ beaker (this is about half a beakerful). Stand the beaker on a large watch glass in a fume cupboard. Pour onto the sugar about 20 cm³ of concentrated sulphuric acid. The sugar will turn yellow, then brown and after about a minute it will start to blacken and a spongy mass of carbon will begin to rise up the beaker and steam will be evolved. The carbon will eventually rise to two or three times the height of the beaker. The steam can be tested with cobalt chloride paper which will go from blue to pink. Sulphur dioxide is also given off and this will turn potassium dichromate paper from orange to blue-green. The beaker becomes very hot. If one drop of water is squirted from a wash bottle onto the outside of the beaker, the drop will steam.

Place about 3 g of blue hydrated copper sulphate on a watch glass and pour onto it about 2 cm³ of concentrated sulphuric acid. Over a period of about three minutes the colour will change to white as the acid dehydrates the salt. Heat is evolved. The change can be reversed by adding water.

Paper and wood shaving can also be dehydrated with sulphuric acid.

Visual tips

A portable fume cupboard gives all-round vision.

Teaching tips

If the students are not familiar with the colours of hydrated and anhydrous copper sulphate, demonstrate the effect of heating blue copper sulphate and then adding water to the white anhydrous salt that is formed.

This demonstration is a spectacular warning of the danger of handling concentrated sulphuric acid. Point out that eye tissue is almost entirely water.

Theory

The reaction is usually written as

$$C_{12}H_{22}O_{11}(s) \rightarrow 12C(s) + 11H_2O(l)$$

but this is an oversimplification. Some of the carbon is oxidised to carbon dioxide and carbon monoxide and some of the sulphuric acid is reduced to sulphur dioxide. There are probably other products.

Further details

See the article The action of concentrated sulphuric acid on sucrose, E. G. Meeks, *Sch. Sci. Rev.*, 1979, **61** (215), 281.

Extensions

Glucose can be dehydrated in the same way, but it is a little slower and there is a longer lag time before the reaction gets underway.

Safety

Wear eye protection.

Work in a fume cupboard because of the gases evolved.

The carbon 'sponge' can be a disposal problem. Place it in its beaker in a large bowl of water and leave for some time to dilute any remaining acid. Small quantities can be broken up with a gloved hand and flushed down the sink. Larger amounts can be placed inside several sealed plastic bags and placed in the dustbin.

It is the responsibility of teachers doing this demonstration to carry out an appropriate risk assessment.

THE ROYAL
SOCIETY OF
CHEMISTRY

56. The density of carbon dioxide

Topic

Gases, properties of carbon dioxide.

Timing

About 2 min.

Level

Lower secondary.

Description

Carbon dioxide is poured onto a burning candle and extinguishes it.

Apparatus

▼   Two 100 cm³ beakers.

▼   Two short pieces of candle (each about 1 cm long).

▼   One 250 cm³ conical flask.

▼   Wooden spills.

▼   If another source of carbon dioxide is not available, one 500 cm³ flask with a side arm or delivery tube fitted with a one-holed rubber bung and a tap funnel to act as a carbon dioxide generator (or *see figure*).

▼   A trough or washing up bowl.

Chemicals

The quantities given are for one demonstration.

▼   Carbon dioxide cylinder with regulator or a few small pieces of dry ice (solid carbon dioxide) or about 10 g of calcium carbonate lumps (marble chips, $CaCO_3$) and about 100 cm³ of 2 mol dm⁻³ **hydrochloric acid** for use in the carbon dioxide generator.

▼   A few cm³ of limewater (a saturated aqueous solution of calcium hydroxide, $Ca(OH)_2$) (optional).

Method

Before the demonstration

Fill the 250 cm³ flask with carbon dioxide from the cylinder or generator by upward displacement of air making sure that the generator and delivery tube are purged of air. Alternatively place a few small lumps of dry ice in the bottom of the flask and leave it for a few minutes to fill with carbon dioxide. This last method has the advantage that the flask replenishes itself automatically with carbon dioxide.

The demonstration

Place the two beakers side by side on the bench and put a short length of candle in each. Light the candles with a spill. They will continue to burn. Pour carbon dioxide from the flask into one of the beakers and the candle will go out while the other continues to burn. Attempts to re-light this candle with a spill will fail and the spill will go out until the carbon dioxide is poured out of the beaker.

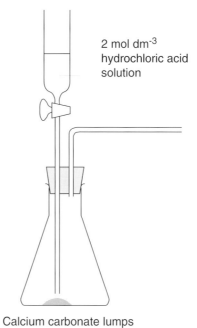

2 mol dm^{-3}
hydrochloric acid
solution

Calcium carbonate lumps

Carbon dioxide generator

Carbon dioxide can be poured into a container which holds a little limewater and the limewater will turn milky.

Visual tips

The demonstration can be scaled up if required.

Teaching tips

Point out the use of carbon dioxide in some types of fire extinguisher.

Theory

Carbon dioxide (relative molecular mass 44) is about one and a half times denser than air (average relative molecular mass about 29).

Safety

Wear eye protection.
It is the responsibility of teachers doing this demonstration to carry out an appropriate risk assessment.

THE ROYAL
SOCIETY OF
CHEMISTRY

57. The enthalpy and entropy changes on the vaporisation of water

Topic

Thermodynamics: latent heat, enthalpy and entropy changes; Trouton's rule.

Timing

About 5 min.

Level

Post-16 but the idea of latent heat of vaporisation could be useful pre-16.

Description

A litre of water is kept boiling for about 2 minutes in an electric kettle and the mass of water boiled away is determined by weighing. The enthalpy and entropy changes of vaporisation can be calculated from the power rating of the kettle.

Apparatus

▼ One domestic electric kettle.

▼ Stopwatch or clock with second hand.

▼ Access to a top pan balance capable of weighing the kettle and about 1 kg of water to the nearest gram.

▼ One 1 dm³ measuring cylinder.

▼ Sticky insulating tape.

Chemicals

▼ Tap water.

Method

Before the demonstration
Check that 900 cm³ of water covers the kettle element and that the balance is capable of weighing the kettle plus 1 dm³ (1 kg) of water. If not, the instructions will need to be adjusted appropriately to take account of the different amount of water used.

The demonstration
Pour 1 dm³ of water into the kettle and bring it to the boil. Switch off and weigh the kettle plus hot water. Switch the kettle back on and start timing. Keep the water boiling for 100 seconds by holding the trip switch down manually or sticking it down with insulating tape. (This may not be necessary on an older type of kettle.) After the 100 seconds is up, switch off the kettle and re-weigh to determine the mass of water that has boiled away. Alternatively, if no suitable balance is available, cool and pour the remaining water into a 1 dm³ measuring cylinder and determine the volume which has boiled away.

Calculate the enthalpy change of vaporisation as follows:

The power rating of the kettle gives the number of kJ supplied per second, hence calculate the number of kJ supplied in 100 seconds to boil away the measured mass

THE ROYAL
SOCIETY OF
CHEMISTRY

of water. Convert this into kJ mol^{-1}. The entropy of vaporisation can now be calculated using the relationship $\Delta S = \Delta H/T$.

Typically a 2.4 kW kettle will boil away 1 g of water per second. This gives a value for H_{vap} of water of 43 kJ mol^{-1} compared with a data book figure of 41.4 kJ mol^{-1}.

Teaching tips

Students may need reminding that the power rating ('wattage') of a kettle is the same as the heat energy supplied per second.

Theory

A value for ΔS_{vap} of water of about 116 J mol^{-1}K^{-1} is obtained (data book figure 108.8 J mol^{-1}K^{-1}). This is unusually large because the hydrogen bonding between water molecules in the liquid state results in a more ordered liquid state than expected and thus a greater increase in disorder on boiling.

Trouton's rule states that the ratio $\Delta H_{vap}/T_b$ is constant. It is obeyed by many liquids. Exceptions include those that form hydrogen bonds. A typical liquid that obeys Trouton's rule has a value of ΔS_{vap} of about 85 J mol^{-1}K^{-1}.

Extensions

An alternative method is to use a small immersion heater instead of a kettle. These can be bought from electrical appliance shops; they are used for making single cups of tea. Their power ratings are less than those of kettles and the experiment will take correspondingly longer.

Other liquids could be tried, but care should be taken over the choice of liquids – with regard to flammability and toxicity of their vapours and compatibility with the material of the kettle. Students could be asked to criticise the design of the experiment and suggest improvements to reduce errors, such as insulating the kettle.

Further details

Students may be impressed at how an apparently obscure piece of information such as an entropy change can be found using simple kitchen apparatus.

Safety

Wear eye protection.

Kettles used in schools should be tested regularly for electrical safety.

It is the responsibility of teachers doing this demonstration to carry out an appropriate risk assessment.

THE ROYAL
SOCIETY OF
CHEMISTRY

58. Catalysts for the decomposition of hydrogen peroxide

Topic

Reaction rates, catalysis, enzymes.

Timing

About 5 min.

Level

Pre-16.

Description

Several measuring cylinders are set up each containing a little washing up liquid and a small amount of a catalyst for the decomposition of hydrogen peroxide. Hydrogen peroxide is poured into the cylinders and a foam rises up the cylinders at a rate that depends on the effectiveness of the catalyst.

Apparatus

▼ Several 250 cm³ measuring cylinders – one for each catalyst to be used.

▼ A large tray to catch any foam that spills over the top of the cylinders.

▼ Stopwatch or clock with second hand.

Chemicals

The quantities given are for one demonstration.

▼ 75 cm³ of **100 volume hydrogen peroxide** solution.

▼ About 0.5 g of powdered manganese(IV) oxide (manganese dioxide, MnO_2).

▼ About 0.5 g of **lead(IV) oxide** (lead dioxide, PbO_2).

▼ About 0.5 g of iron(III) oxide (red iron oxide, Fe_2O_3).

▼ A small piece (about 1 cm³) of potato.

▼ A small piece (about 1 cm³) of liver.

Method

Before the demonstration

Line up five 250 cm³ measuring cylinders in a tray. Add 75 cm³ of water to the 75 cm³ of 100 volume hydrogen peroxide solution to make 150 cm³ of 50 volume solution.

The demonstration

Place about 1 cm³ of washing up liquid into each of the measuring cylinders. To each one add the amount of catalyst specified above. Then add 25 cm³ of 50 volume hydrogen peroxide solution to each cylinder. The addition of the catalyst to each cylinder should be done as nearly simultaneously as possible – using two assistants will help. Start timing. Foam will rise up the cylinders. The lead dioxide will probably

THE ROYAL
SOCIETY OF
CHEMISTRY

be fastest, followed by manganese dioxide and liver. Potato will be much slower and the iron oxide will barely produce any foam. This order could be affected by the surface areas of the powders. Time how long each foam takes to rise to the top (or other marked point) of the cylinder. The foam from the first three cylinders will probably overflow considerably.

Place a glowing spill in the foam; it will re-light confirming that the gas produced is oxygen.

Teaching tips

Some students may believe that the catalysts – especially the oxides – are reactants because hydrogen peroxide is not noticeably decomposing at room temperature. The teacher could point out the venting cap on the peroxide bottle as an indication of continuous slow decomposition. Alternatively s/he could heat a little hydrogen peroxide in a conical flask with a bung and delivery tube, collect the gas over water in a test-tube and test it with a glowing spill to confirm that it is oxygen. This shows that no other reactant is needed to decompose hydrogen peroxide.

NB: Simply heating 50 volume hydrogen peroxide in a test-tube will not suceed in demonstrating that oxygen is produced. The steam produced will tend to put out a glowing spill. Collecting the gas over water has the effect of condensing the steam. It is also possible to 'cheat' by dusting a beaker with a tiny, almost imperceptible, amount of manganese dioxide prior to the demonstration and pouring hydrogen peroxide into it. Bubbles of oxygen will be formed in the beaker.

Theory

The reaction is :

$$2H_2O_2(aq) \rightarrow 2H_2O(l) + O_2(g)$$

This is catalysed by a variety of transition metal compounds and also by peroxidase enzymes found in many living things.

Extensions

Repeat the experiment but heat the liver and the potato pieces for about five minutes in boiling water before use. There will be almost no catalytic effect, confirming that the catalyst in these cases is an enzyme that is denatured by heat.

Investigate the effect of using lumpy or powdered manganese dioxide. The powdered oxide will be more effective because of its greater surface area.

Try using other metal oxides or iron filings as catalysts.

Animal blood may be used instead of liver if local regulations allow this.

One teacher suggested measuring the height of the foam over suitable time intervals and plotting a graph.

Further details

The experiment can be done with 20 volume hydrogen peroxide, but is less spectacular. It is, however, easier to time.

It has been suggested that manganese dioxide is not in fact the catalyst for this reaction, but that the catalysts are traces of other oxides found on the surface of manganese dioxide.

Safety

Wear eye protection.

Used liver should be wrapped up in paper and placed in the dustbin.

It is the responsibility of teachers doing this demonstration to carry out an appropriate risk assessment.

THE ROYAL
SOCIETY OF
CHEMISTRY

59. Estimating the concentration of domestic bleach

Topic

Everyday chemistry, reactions of chlorine compounds, redox reactions.

Timing

About 5 min.

Level

Any, but post-16 to fully appreciate the calculations.

Description

Excess hydrogen peroxide is added to household bleach and the volume of oxygen produced is measured. The concentration of sodium chlorate(I) in the bleach can be calculated.

Apparatus

▼   One 250 cm³ Buchner flask with a one-holed rubber bung to fit.

▼   One 10 cm³ plastic syringe.

▼   About 30 cm of rubber tubing to fit the side arm of the Buchner flask.

▼   One 100 cm³ measuring cylinder.

▼   Trough or washing up bowl.

▼   One 5 cm³ pipette and filler.

Chemicals

The quantities given are for one demonstration.

▼   5 cm³ of household **bleach**.

▼   10 cm³ of 20 volume hydrogen peroxide solution.

Method

Before the demonstration

Set up the apparatus shown or its equivalent. For example a conical flask with a two-holed bung and a delivery tube could be used instead of the Buchner flask. A gas syringe could be used instead of collecting the oxygen over water. A hypodermic needle could be used to inject the hydrogen peroxide through a septum cap.

The demonstration

Use the pipette to measure 5 cm³ of household bleach into the flask and replace the bung. Fill the syringe with 10 cm³ of 20 volume hydrogen peroxide and fit its nozzle into the rubber bung. Fill the measuring cylinder with water. Squirt the peroxide into the bleach. The solutions will react and oxygen will be given off and collected in the measuring cylinder. Shake to ensure complete mixing and take a reading of the volume of the gas collected when gas has stopped being given off. Calculate the percentage of sodium chlorate(I) in the bleach.

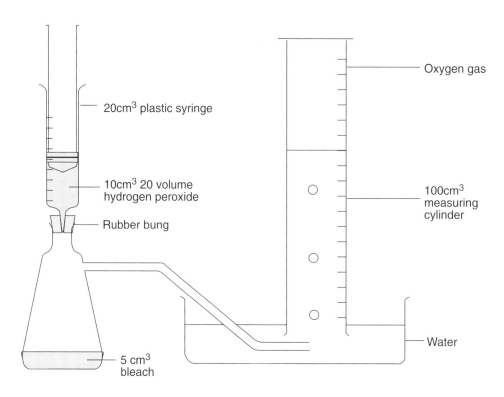

Generating oxygen

Teaching tips

For classes that might have difficulty with the calculations, the measuring cylinder could be directly calibrated in % sodium hypochlorite before the demonstration *eg* 90 cm³ could be marked 5 % *etc*. See the calculations below.

If a list of the prices of different brands of bleach was compiled, a value for money figure could be worked out such as 'grams of hypochlorite per penny'. A spreadsheet would be ideal for this type of calculation.

Theory

The active ingredient of household bleach is sodium chlorate(I) (sodium hypochlorite). This reacts with hydrogen peroxide:

$$NaOCl(aq) + H_2O_2(aq) \rightarrow NaCl(aq) + H_2O(l) + O_2(g)$$

So one mole of oxygen is equivalent to one mole (74.5 g) of sodium hypochlorite and the percentage of sodium hypochlorite in the original bleach solution can be calculated as follows:

If the total volume of gas in the measuring cylinder is V cm³, then the volume of oxygen produced will be V–10 cm³ because the hydrogen peroxide will displace 10 cm³ of air. At room temperature this is approximately (V–10)/24 000 mole. So 5 cm³ of bleach solution contained (V–10)/24 000 mole of sodium hypochlorite which is [(V–10)/24 000] × 74.5 g. So in 100 cm³ of solution there would be 20 × [(V–10)/24 000] × 74.5, *ie* 0.062(V–10) % sodium hypochlorite.

So normal household bleach (concentration of less than 5 %) should give a maximum of 90 cm³ of gas in total.

The density of bleach is about 1.2 g cm⁻³ so the w/v percentage could be converted into w/w if required.

THE ROYAL
SOCIETY OF
CHEMISTRY

Extensions

Some brands of thickened bleach seem to give lower results than expected. It is best to add water to these after pipetting 5 cm³ of bleach into the flask. Students may need help to appreciate that this dilution does not affect the amount of bleach present. Results could be compared with those obtained by adding excess potassium iodide to the bleach and titrating the liberated iodine with standard sodium thiosulphate.

Further details

This demonstration has been adapted from an idea by A. Jackson and J. McGregor, *Sch. Sci. Rev.,* **62** (219), 1980, 322. Their calculations are done in terms of 'available chlorine'.

50 volume hydrogen peroxide can also be used in which case the reaction goes faster and it is easier to see when gas evolution has ceased.

Safety

Wear eye protection.

It is the responsibility of teachers doing this demonstration to carry out an appropriate risk assessment.

THE ROYAL
SOCIETY OF
CHEMISTRY

60. The reaction of ethyne with chlorine

Topic

Organic chemistry, reactions of alkynes or general interest.

Timing

About 5 min.

Level

Post-16, or any for general interest.

Description

Calcium carbide reacts with dilute sulphuric acid to produce ethyne gas. Household bleach is squirted into the acid and forms chlorine. The ethyne and chlorine react together vigorously giving flames and producing a black soot of carbon.

Apparatus

▼   One 250 cm^3 beaker.

▼   One dropping pipette.

▼   Access to a fume cupboard (optional).

Chemicals

The quantities given are for one demonstration.

▼   About 50 cm^3 of approximately 2 mol dm^{-3} **sulphuric acid**.

▼   One lump (about the size of a pea) of **calcium carbide** (calcium dicarbide, CaC$_2$).

▼   About 5 cm^3 of household **bleach** – an approximately 5 % solution of sodium chlorate(I) (sodium hypochlorite, NaOCl) and sodium chloride.

Method

The demonstration

Place 50 cm^3 of 2 mol dm^{-3} sulphuric acid in a 250 cm^3 beaker. Add one pea-sized lump of calcium carbide. This will react to give off bubbles of ethyne (acetylene, C$_2$H$_2$). Now add about 1 cm^3 of domestic bleach to the acid using a dropping pipette. Chlorine gas is evolved. Within a few seconds (or possibly immediately) the two gases will react giving a yellow flame and black soot. Intermittent flames will continue for about a minute. More bleach or carbide may be added as appropriate to continue the reaction. Some teachers may prefer to perform this demonstration in the fume cupboard.

Teaching tips

The reaction of calcium carbide with water to form ethyne is still used occasionally by cavers to produce the yellow flame in 'carbide lamps'.
Students will probably predict that the reaction of ethyne with chlorine will be addition across the double bond:

$$C_2H_2(g) + Cl_2(g) \rightarrow CHCl=CHCl(l)$$

THE ROYAL
SOCIETY OF
CHEMISTRY

or

$$C_2H_2(g) + 2Cl_2(g) \rightarrow CHCl_2\text{--}CHCl_2(l)$$

These take place in solution, but the main reaction that actually takes place in the gas phase appears to be a radical reaction as shown below.

Theory

The reactions are:

$$CaC_2(s) + H_2SO_4(aq) \rightarrow C_2H_2(g) + CaSO_4(s)$$

$$ClO^-(aq) + Cl^-(aq) + 2H^+(aq) \rightarrow Cl_2(g) + H_2O(l)$$

$$C_2H_2(g) + Cl_2(g) \rightarrow 2C(s) + 2HCl(g)$$

The intermittent reaction seems to be caused by a build up of ethyne followed by reaction in which it is all used up . More ethyne is evolved and it again builds up to a critical value and so on.

Extensions

Ethyne can be made by reacting calcium carbide with water in a conical flask with a delivery tube and collected over water in small test-tubes. It can be shown to undergo the expected reactions:

▼   it decolorises bromine water;

▼   it reacts with solutions of potassium manganate(VII) (turning acidic solutions brown (manganese dioxide) and alkaline solutions green (potassium manganate (VI)); and

▼   It burns with a smoky flame (ensure that all the air has been swept out of the apparatus before lighting the gas).

Safety

Wear eye protection.
    It is the responsibility of teachers doing this demonstration to carry out an appropriate risk assessment.

61. Identifying the products of combustion

Topic

Combustion reactions.

Timing

About 5 min.

Level

Introductory chemistry.

Description

A candle is burned and a filter pump is used to draw the gaseous combustion products over a piece of cobalt chloride paper and through limewater to identify water and carbon dioxide respectively.

Apparatus

▼ One glass funnel, about 6 cm in diameter.

▼ Two boiling tubes.

▼ Two two-holed rubber bungs to fit the boiling tubes fitted with one long and one short piece of glass tubing.

▼ One filter pump.

▼ Glass or plastic tubing for connections.

Chemicals

The quantities given are for one demonstration.

▼ One candle.

▼ One piece of blue cobalt chloride paper.

▼ About 20 cm³ of limewater (a saturated solution of calcium hydroxide, $Ca(OH)_2$).

Method

Before the demonstration

Assemble the apparatus shown in *Fig. 1*. Care should be taken with the right angle bend which is connected to the funnel. If this is made of flexible tubing, it can get hot and melt. Ideally, the glass stem of the funnel should be bent into a right-angle. Alternatively, join a standard funnel onto a right angled piece of glass tubing using epoxy resin. A more temporary arrangement is to slide one arm of a right-angled piece of glass tubing inside the stem of the funnel and seal the join on the outside with a piece of flexible tubing *(Fig. 2)*.

The demonstration

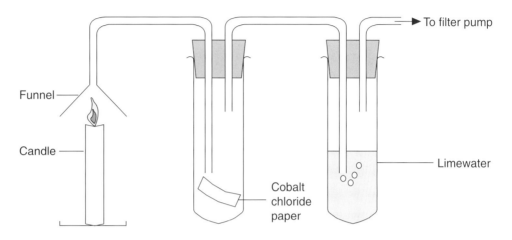

Fig. 1 Apparatus to identify the products of combustion

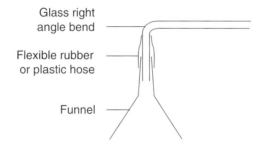

Fig. 2 Temporary right-angled bend

Place a piece of blue cobalt chloride paper into the first boiling tube and half fill the second boiling tube with limewater. Turn on the filter pump so that a gentle stream of air is drawn through the apparatus. Light the candle and leave for a few minutes until the cobalt chloride paper turns pink and the limewater goes milky, indicating the presence of water and carbon dioxide respectively.

Visual tips

The apparatus can be left running for some time and students can file past it in small groups to see it more closely.

Teaching tips

If students are not familiar with the cobalt chloride paper and limewater tests, demonstrate these separately (perhaps while waiting for changes to occur in the main demonstration). Ask the students to explain why it is important that the combustion products are drawn over the cobalt chloride paper before being bubbled through the limewater.

Some students may know that air contains both water vapour and carbon dioxide. To show that the changes observed are not due to these alone, repeat the experiment without the candle and note how much longer it takes for any changes to be observed.

THE ROYAL
SOCIETY OF
CHEMISTRY

The identification of carbon dioxide could lead to a discussion of the role of this gas in the greenhouse effect.

Theory

The precipitate formed in the reaction of carbon dioxide with limewater is calcium carbonate:

$$Ca(OH)_2(aq) + CO_2(g) \rightarrow CaCO_3(s) + H_2O(l).$$

If carbon dioxide is bubbled through for longer, the precipitate may re-dissolve as calcium hydrogencarbonate (calcium bicarbonate):

$$CaCO_3(s) + H_2O(l) + CO_2(g) \rightarrow Ca(HCO_3)_2(aq).$$

Extensions

Other fuels can be used, liquids being burned in a spirit burner.

Further details

If the tube that normally contains cobalt chloride is left empty, but is immersed in a beaker of ice water, it may be possible to collect enough water for the boiling point and freezing point to be measured if the demonstration can be left long enough.

Safety

Wear eye protection.

It is the responsibility of teachers doing this demonstration to carry out an appropriate risk assessment.

THE ROYAL
SOCIETY OF
CHEMISTRY

62. The spontaneous combustion of iron

Topic

Reactions of metals, oxidation, effect of surface area on reaction rate, or general interest.

Timing

About 2 min.

Level

Introductory chemistry.

Description

Iron(II) ethanedioate decomposes on heating to give carbon dioxide, carbon monoxide and finely divided iron(II) oxide and iron. On exposure to the air this is oxidised rapidly and exothermically to iron(III) oxide giving a 'sparkler' effect.

Apparatus

▼   One boiling tube.

▼   Heat-proof mat.

▼   Bunsen burner.

▼   Test-tube holder.

▼   A little mineral wool.

▼   Glass rod (optional).

Chemicals

The quantities given are for one demonstration.

▼   About 2 g of **iron(II) ethanedioate** (iron(II) ethanedioate-2-water, ferrous oxalate, $FeC_2O_4.2H_2O$).

▼   A little limewater (saturated calcium hydroxide solution, $Ca(OH)_2(aq)$) (optional).

Method

The demonstration

Place about 2 g of iron(II) ethanedioate in a boiling tube, plug the top loosely with a little mineral wool and heat the yellow solid over a roaring Bunsen flame. The mineral wool is to stop any dust escaping when the salt is heated. Water of crystallisation is given off and the steam condenses around the rim of the boiling tube. The solid 'seethes' as carbon dioxide and carbon monoxide are given off and the yellow powder turns grey. Continue heating, shaking occasionally, until all the yellow colour is gone and the powder has stopped 'seething'. At this stage it is possible to test the gas in the tube by holding a drop of limewater on a glass rod at the mouth of the tube. It will go milky showing that carbon dioxide is present. Alternatively use a delivery tube to bubble the gas through a beaker of limewater. The gas does not appear to burn.

   Now remove the mineral wool plug and pour the solid product onto a heat-proof

mat from a height of about half a metre. Freed of its protective blanket of carbon
dioxide and on contact with the air, the finely divided iron and iron(II) oxide mixture
is rapidly and exothermically oxidised to red iron(III) oxide and the powder glows red
hot.

   The finely divided mixture is referred to as pyrophoric *ie* capable of burning
spontaneously on exposure to air.

Visual tips

The pouring would look most spectacular in a darkened room.

Teaching tips

This is a good illustration of the effect of surface area on reaction rates; the rate of
reaction can be compared with that of rusting and also of sprinkling iron filings into a
Bunsen flame. Firework sparklers contain finely divided iron.

Theory

The reactions can be written as

$$FeC_2O_4(s) \rightarrow Fe(s) + 2CO_2(g)$$

and

$$FeC_2O_4(s) \rightarrow FeO(s) + CO(g) + CO_2(g)$$

$$4Fe(s) + 3O_2(g) \rightarrow 2Fe_2O_3(s)$$

and

$$4FeO(s) + O_2(g) \rightarrow 2Fe_2O_3(s)$$

Extensions

Lead(II) 2,3-dihydroxybutanedioate (lead(II) tartrate) and tin(II) ethanedioate (tin(II)
oxalate) are reported to undergo similar decompositions to produce pyrophoric
products.

Further details

If the iron ethanedioate is heated in a suitable vial, the pyrophoric product can be
sealed inside with a glass-blowing torch. The vial can later be broken open and the
oxidation of the powder demonstrated.

   Iron(II) ethanedioate can be prepared by mixing solutions of iron(II) sulphate and
sodium ethanedioate and filtering and drying the resulting precipitate.

Safety

Wear eye protection.

   The room should be well ventilated because of the carbon monoxide generated.

   It is the responsibility of teachers doing this demonstration to carry out an
appropriate risk assessment.

THE ROYAL
SOCIETY OF
CHEMISTRY

63. The thermal decomposition of nitrates – 'magic writing'

Topic

Thermal decomposition of nitrates, general interest.

Timing

About 5 min.

Level

Lower secondary.

Description

A message is written on filter paper with a solution of sodium nitrate and is then dried. Applying a glowing taper to the start of the message makes the treated paper smoulder and the message is revealed as the glow spreads its way through the treated paper only.

Apparatus

▼   Filter or blotting paper sheets – as large as possible.

▼   Wooden taper.

▼   Bunsen burner or hair-drier.

▼   Small paint brush.

Chemicals

The quantities given are for one demonstration.

▼   About 10 g of **sodium nitrate** (sodium nitrate(V), $NaNO_3$).

Method

Before the demonstration
Make a saturated solution of sodium nitrate by adding about 10 g of solid to 10 cm^3 of water and stirring. Using a small paintbrush (or a length of wooden taper), write a message on the filter paper. Use joined up writing! Dry the message using a hair-drier or by holding the paper well above a Bunsen flame. The message will be virtually invisible, so mark the start of it with a light pencil mark.

The demonstration
Pin up the filter paper in the sight of the audience. Apply a glowing taper to the start of the message until the treated paper starts to glow and char. Remove the taper and watch as the glow and charring works its way along the message, leaving the untreated paper untouched.

Teaching tips

This demonstration could be used to introduce the fire triangle: fuel, heat and oxygen. With older students it could be used to revise the equations for the decomposition of **nitrates**.

Theory

The reaction that occurs is:

$$2NaNO_3(s) \rightarrow 2NaNO_2(s) + O_2(g)$$

The oxygen produced is sufficient to keep the treated paper smouldering while the untreated paper does not burn.

Extensions

Try other metal nitrates (see below).
What effect does the nitrate concentration have?

Further details

Potassium nitrate works in the same way as sodium nitrate.
Lithium nitrate also works although it decomposes slightly differently due to the higher charge density on the lithium ion

$$4LiNO_3(s) \rightarrow 2Li_2O(s) + 4NO_2(g) + O_2(g)$$

Lead nitrate will also work:

$$2Pb(NO_3)_2(s) \rightarrow 2PbO(s) + 4NO_2(g) + O_2(g)$$

Ammonium nitrate does not work because it does not give off oxygen as it decomposes:

$$NH_4NO_3(s) \rightarrow N_2O(g) + 2H_2O(l)$$

Although nitrogen(I) oxide (N_2O) will itself decompose to give oxygen, there is presumably either insufficient N_2O to keep the paper smouldering or the temperature is too low to bring about decomposition.

Safety

Wear eye protection.
It is the responsibility of teachers doing this demonstration to carry out an appropriate risk assessment.

THE ROYAL
SOCIETY OF
CHEMISTRY

64. Making nylon – the 'nylon rope trick'

Topic

Polymerisation.

Timing

About 5 min.

Level

Pre-16 or post-16, depending on the sophistication of theoretical treatment.

Description

A solution of decanedioyl dichloride in cyclohexane is floated on an aqueous solution of 1,6-diaminohexane. Nylon forms at the interface and can be pulled out as fast as it is produced forming a long thread – the 'nylon rope'.

Apparatus

▼   One 25 cm³ beaker.

▼   A pair of tweezers.

▼   Retort stand with boss and clamp.

Chemicals

The quantities given are for one demonstration.

▼   2.2 g of **1,6-diaminohexane** (hexamethylene diamine, hexane-1,6-diamine, $H_2N(CH_2)_6NH_2$).

▼   1.5 g of **decanedioyl dichloride** (sebacoyl chloride, $ClOC(CH_2)_8COCl$).

▼   50 cm³ of **cyclohexane**.

▼   50 cm³ of deionised water.

Method

Before the demonstration

Make up a solution of 2.2 g of 1,6-diaminohexane in 50 cm³ of deionised water. This solution is approximately 0.4 mol dm⁻³.

Make up a solution of 1.5 g of decanedioyl dichloride in 50 cm³ of cyclohexane. This solution is approximately 0.15 mol dm⁻³.

The demonstration

Pour 5 cm³ of the aqueous diamine solution into a 25 cm³ beaker. Carefully pour 5 cm³ of the cyclohexane solution of the acid chloride on top of the first solution so that mixing is minimised. Do this by pouring the second solution down the wall of the beaker or pour it down a glass rod. The cyclohexane will float on top of the water without mixing. Place the beaker below a stand and clamp as shown (see figure). A greyish film of nylon will form at the interface. Pick up a little of this with a pair of tweezers and lift it slowly and gently from the beaker. It should draw up behind it a thread of nylon. Pull this over the rod of the clamp so that this acts as a pulley. Continue pulling the nylon thread at a rate of about half a metre per second. It should be possible to pull out several metres. Take care, the thread will be coated with unreacted monomer and may in fact be a narrow, hollow tube filled with monomer solution. Wearing disposable gloves is a wise precaution.

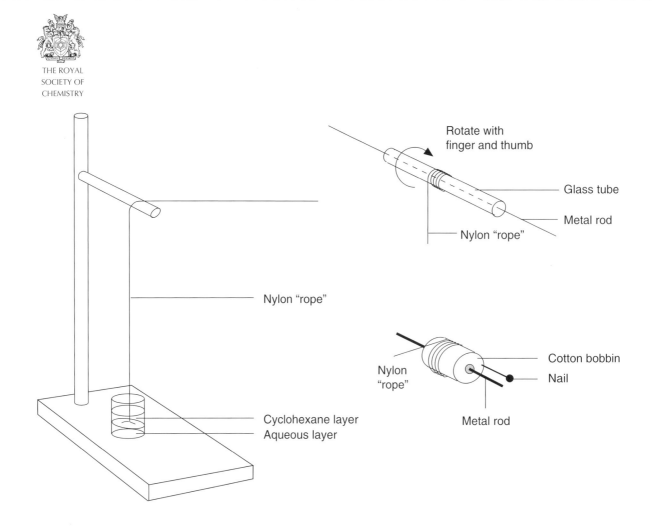

THE ROYAL
SOCIETY OF
CHEMISTRY

The nylon rope trick

Visual tips

The beaker is rather small so allow the audience as close as possible consistent with comfort and safety.

Teaching tips

Point out that this demonstration is different from the industrial method of making nylon which takes place at a higher temperature. Molten nylon is then forced through multi-holed 'spinnerets' to form the fibres.

Theory

The reaction is a condensation polymerisation

$$nH_2N(CH_2)_6NH_2 + nClOC(CH_2)_8COCl \rightarrow H_2N\left[(CH_2)_6NHCO(CH_2)_8\right]_n COCl + nHCl$$

The nylon formed is nylon 6–10 so called because of the lengths of the carbon chains of the monomers. Nylon 6–6 can be made using hexanedioyl dichloride (adipoyl chloride).

The diamine is present in excess to react with the hydrogen chloride that is eliminated. An alternative procedure is to use the stoichiometric quantity of diamine dissolved in excess sodium hydroxide solution.

THE ROYAL
SOCIETY OF
CHEMISTRY

Extensions

There are many ways of conveniently winding the nylon thread – for example using a windlass improvised from a cotton bobbin or a short length of glass tube slid over the rod of a clamp *(see Fig)*.

Further details

This demonstration has been described in many sources using chlorinated solvents for the acid chloride. These are no longer considered safe and will soon become unavailable. Cyclohexane is less dense than water whereas chlorinated solvents are denser. The layers are therefore inverted compared with the old method.

Cyclohexane is preferred to hexane as it is less harmful.

Hexanedioyl dichloride (adipoyl chloride) can be used as an alternative to decanedioyl dichloride, but it does not keep as well.

Decanedioyl dichloride reacts with moisture in the air to produce decanedioic acid which forms nylon much less readily than the acid chloride. Ensure that the bottle is re-stoppered carefully after opening and consider storing it in a desiccator. The dichloride is also available in 5 cm^3 sealed ampoules. The cyclohexane solution will still make nylon for a couple of days after being made up even if left unstoppered. A solution kept in a stoppered bottle is still usable after two weeks. The solution can be stored over anhydrous sodium sulphate or calcium chloride to keep it dry.

Solid 1,6-diaminohexane can be difficult to get out of the bottle. The easiest way to manipulate it is to heat the bottle gently in warm water until it melts at 42 °C and dispense the liquid using a dropping pipette.

Safety

Wear eye protection.

Dispose of the mixture as follows:

First shake the reaction to mix the two layers. A lump of nylon will be produced which can be removed with tweezers, rinsed well with water, and disposed as solid waste. Failure to do this may result in the polymerisation taking place in the sink, leading to a blockage. The remaining liquids can be mixed with detergent and washed down the sink.

It is the responsibility of teachers doing this demonstration to carry out an appropriate risk assessment.

65. Diffusion of gases – ammonia and hydrogen chloride

Topic

Gases, diffusion, kinetic theory.

Timing

About 5 min.

Level

Lower secondary.

Description

Solutions of concentrated ammonia and concentrated hydrochloric acid are soaked into cotton wool plugs and placed in either end of a glass tube. After about a minute, the gases diffuse and meet near the middle of the tube where they react to form a white ring of solid ammonium chloride.

Apparatus

▼   A lengh of glass tube about half a metre long with an inside diameter of about 2 cm.

▼   Two retort stands with bosses and clamps.

▼   Two small wads of cotton wool.

▼   Access to a fume cupboard (optional).

Chemicals

The quantities given are for one demonstration.

▼   A few cm³ of **concentrated hydrochloric acid**.

▼   A few cm³ of **880 ammonia solution**.

These solutions are more easily manipulated if they are in small bottles rather than Winchesters.

Method

Before the demonstration
Ensure that the glass tube is clean and dry inside. It can be dried by pushing a wad of cotton wool soaked in propanone through it and leaving it for a few minutes. Clamp the tube in view of the audience. The demonstration should be done in a fume cupboard or a well-ventilated room.

The demonstration
Working in a well ventilated room (or a fume cupboard) and wearing plastic gloves, open the ammonia and hydrochloric acid bottles and blow the fumes from one over the other. Note the white clouds of ammonium chloride that form. Now take two small wads of cotton wool which will fit into the end of the glass tube. Place one over the neck of the ammonia bottle and invert the bottle so that a little ammonia soaks into the cotton wool but the whole of the wad is not soaked. Place the moistened end of the cotton wool in one end of the glass tube. Quickly repeat this

THE ROYAL
SOCIETY OF
CHEMISTRY

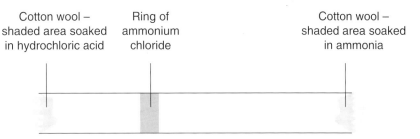

Cotton wool –	Ring of	Cotton wool –
shaded area soaked	ammonium	shaded area soaked
in hydrochloric acid	chloride	in ammonia

Diffusion of ammonia and hydrogen chloride

procedure with the second wad of cotton wool but using hydrochloric acid and placing it in the other end of the glass tube. Some teachers may wish to stopper the ends of the tube after inserting the cotton wool. Watch the tube. After about one minute a white ring of solid ammonium chloride will form in the tube. This will be closer to the hydrochloric acid end of the tube because hydrogen chloride molecules have about twice the mass of ammonia molecules and therefore diffuse more slowly. The rate of diffusion is inversely proportional to the square root of the relative molecular mass of the gas. The exact time for the ring to form will depend on the dimensions of the tube, the temperature and how well soaked the cotton wool wads are.

Visual tips

A dark background is better than a light one for seeing the ring.

Teaching tips

Explain that the purpose of the tube is to eliminate air currents and to see if the molecules can move on their own. Point out that the molecules follow a zig-zag 'drunkards walk' type of path as they collide with molecules of air and that the time taken for the ring to form is not a good indication of the speeds of the molecules, which are of the order of hundreds of metres per second.

Theory

The reaction is:

$$NH_3(g) + HCl(g) \rightarrow NH_4Cl(s)$$

Further details

An alternative way of doing this demonstration is to place a cotton wool wad soaked in an indicator at one end of the tube and ammonia or hydrochloric acid at the other. Watch for the indicator to change colour.

One teacher suggested placing a strip of moistened indicator paper along the length of the tube so that the progress of the diffusion can be monitored continuously.

Safety

Wear eye protection.

It is the responsibility of teachers doing this demonstration to carry out an appropriate risk assessment.

THE ROYAL
SOCIETY OF
CHEMISTRY

66. Water as the product of burning hydrogen

Topic

Combustion reactions.

Timing

About 5 min to set up, about one hour to collect sufficient water to test its boiling and freezing points.

Level

Lower secondary.

Description

Hydrogen is burned at a jet which plays on a cold finger type of condenser. Water is collected and can be tested for using cobalt chloride paper and by measuring its boiling and freezing points.

Apparatus

▼ One 250 cm³ conical or round-bottomed flask fitted with a two-holed bung with two glass tubes as shown in *Fig. 1*.

▼ Two lengths of rubber tube to connect the condenser to the tap and to the sink.

▼ A length of rubber tubing to connect the hydrogen cylinder to the syringe.

▼ One 1 cm³ plastic hypodermic syringe and needle.

▼ One 100 cm³ beaker.

▼ Test-tube and holder.

▼ Bunsen burner.

▼ Thermometer (–10 to 110 °C).

▼ Access to freezer.

▼ Safety screen.

Chemicals

The quantities given are for one demonstration.

▼ **Hydrogen** cylinder with valve gear.

▼ About 250 cm³ of ice/salt freezing mixture (optional, if freezer is not available). Mix roughly four parts by volume of crushed ice to one part common salt (sodium chloride).

▼ One piece of blue cobalt chloride paper and/or a little anhydrous **copper sulphate**.

Method

Before the demonstration

Set up the cold finger condenser as shown in *Fig. 1*. The water should enter through

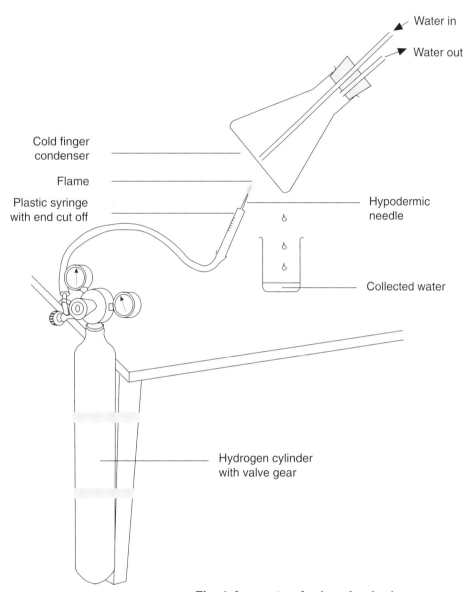

Fig. 1 Apparatus for burning hydrogen

the long glass tube, which should reach almost to the bottom of the flask, and leave through the short one. Clamp it at about 45° to the horizontal and turn on a steady flow of water. Dry the outside of the flask with a paper tissue.

Remove the plunger from a 1 cm³ plastic hypodermic syringe and cut off the end of the barrel so that a length of rubber tube can be fitted over the cut end to connect it to the hydrogen cylinder (*Fig. 2*).

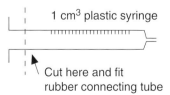

Fig. 2 Hydrogen delivery nozzle

THE ROYAL
SOCIETY OF
CHEMISTRY

The demonstration

Put a safety screen between the apparatus and the audience. Turn on the hydrogen for a few seconds to displace air from the connecting tube. Place the hypodermic needle on the end of the syringe and clamp the syringe below the cold finger condenser with the end of the hypodermic about 2 cm from the flask. Place a beaker below the lowest point of the flask to collect the water drips. Turn on a gentle flow of hydrogen for a few seconds. Collect a small test-tube full of the gas emerging from the needle by downward displacement of air and light it with a taper. If it burns quietly rather than 'popping', the apparatus is free of air and the hydrogen may be lit safely at the needle. Adjust the flame until it is about 1 cm long and it plays on the bottom of the cold finger. Condensation will form on the flask. This can be tested with blue cobalt chloride paper to confirm that it is water. After a few minutes water droplets will begin to drip into the beaker. Now leave the apparatus until about 10 cm^3 of water has been collected, enough to measure its boiling and freezing points. This will take about an hour. During this time the apparatus will need to be watched to ensure that the flame is not blown out.

Test the boiling point of the water by clamping a test-tube containing about 5 cm^3 of the water over a Bunsen burner and holding a thermometer in the steam. It should read close to 100 °C depending on atmospheric pressure.

Test the freezing point by placing about 4 cm^3 of the water in a test-tube. Place a thermometer bulb into the water and place the whole assembly in the freezer (or in ice/salt mixture) until it freezes. Remove from the freezer and allow it to warm up. Note the temperature when both ice and water are present. It should be about 0 °C.

If desired, test the remaining water by placing a few drops onto a little anhydrous copper sulphate on a watch glass. This will turn blue and heat will be evolved.

Visual tips

During the hour or so in which water is being collected, students could file past the apparatus to see it at close quarters.

Teaching tips

Explain that the water inside the cold finger is to keep the flask cool and is separate from the water collecting on the outside.

If the mains water is very cold and the room humid, atmospheric water vapour may condense on the cold finger. If this occurs, it will be necessary to set up two cold fingers, one as a control, to show that there is more condensation on the one with the flame.

Make sure the students are aware of the cobalt chloride and/or anhydrous copper sulphate tests for water and demonstrate them with tap water if necessary.

Because there is an hour or so with little to see during this demonstration, the teacher should have some other activity ready for the students during this period.

Further details

It is not recommended to try this experiment with a source of hydrogen other than from a hydrogen cylinder.

Hydrogen is a possible fuel of the future partly because its only combustion product is water (although some nitrogen oxides will also be formed in the flame).

Safety

Wear eye protection.

Use a safety screen to protect yourself and the pupils while lighting the gas.

It is the responsibility of teachers doing this demonstration to carry out an appropriate risk assessment.

THE ROYAL
SOCIETY OF
CHEMISTRY

67. The greenhouse effect – 1

Topic

Environmental chemistry.

Timing

About 30 min.

Level

Any.

Description

The 'greenhouse effect' in the Earth's atmosphere is caused by a number of gases that behave in a similar way to glass in a greenhouse. In the demonstration, three thermometers are clamped close to a photoflood bulb and their temperatures monitored regularly. One is clamped in the air, one is enclosed in a plastic pop bottle, and one enclosed in a pop bottle one half of which has been painted with matt black paint. The final steady temperatures obtained are in the order 'bare' thermometer (lowest), thermometer in unpainted bottle, thermometer in painted bottle (highest).

Apparatus

▼   Two 1 dm³ plastic fizzy drinks bottles with two-holed rubber bungs to fit.

▼   Three mercury-in-glass thermometers (0–100 °C).

▼   One 275 W photoflood light bulb (obtainable from photographic shops) with a plain bulb holder (ie without a shade).

▼   Clock with second hand.

▼   Three pieces of lead foil about 3 cm x 2 cm.

▼   A little matt black paint such as blackboard paint.

Method

Before the demonstration

Check that all three thermometers give the same reading in the same surroundings. Clean and dry the bottles. Cut three identical pieces of lead foil and fold them round the bulbs of the thermometers to form 'flags' (*Fig. 1*). These absorb the light energy and radiate it as heat, simulating the Earth's surface. Ensure that the thermometers will still fit through the openings in the bottles when the lead 'flags' are fitted. Paint half of one of the bottles with matt black paint as shown in *Fig. 2*. Fit two of the thermometers through the bungs ensuring that it is possible to read their scales from room temperature upwards. Place the bungs holding the thermometers into the two pop bottles.

The demonstration

Stand the photoflood bulb in its holder on the bench. Clamp the three thermometers (two of them inside their bottles) so that they are about 25 cm from the bulb. The actual distance is not critical, but it is important that all three distances are the same.

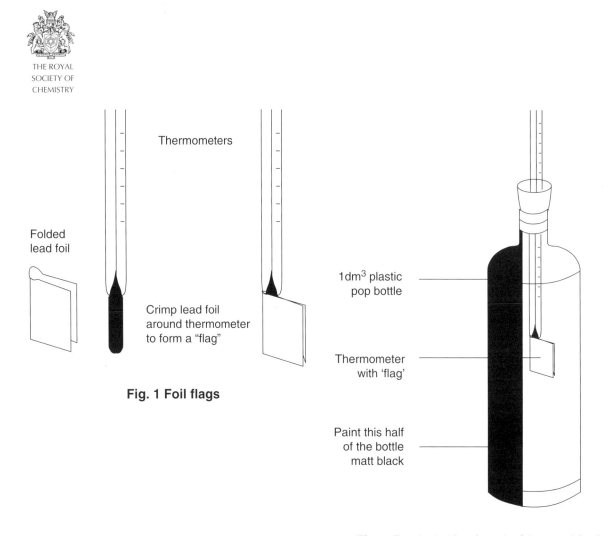

Folded
lead foil

Thermometers

Crimp lead foil
around thermometer
to form a "flag"

Fig. 1 Foil flags

1dm³ plastic
pop bottle

Thermometer
with 'flag'

Paint this half
of the bottle
matt black

Fig. 2 Bottle half-painted with matt black

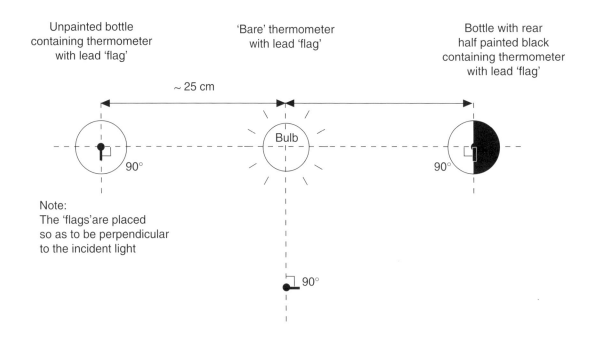

Unpainted bottle
containing thermometer
with lead 'flag'

'Bare' thermometer
with lead 'flag'

Bottle with rear
half painted black
containing thermometer
with lead 'flag'

~ 25 cm

Bulb

90°   90°

Note:
The 'flags'are placed
so as to be perpendicular
to the incident light

90°

Fig. 3 Top view of apparatus

THE ROYAL
SOCIETY OF
CHEMISTRY

A previously prepared paper template on which the positions of the apparatus are marked will help when setting this up in front of a class. The bulbs of the thermometers should be at the same level as the photoflood bulb and the lead 'flags' should be perpendicular to the incident light (*Fig. 3*). Allow the thermometers to adjust to room temperature and take a reading of each. Switch on the photoflood bulb, start the clock and take a reading of each thermometer every minute for about 15 minutes. The temperatures of each will rise and gradually level off to a steady reading. Typically the 'bare' thermometer's reading will rise by 5 °C, the one in the clear bottle by 8 °C and the one in the half-blackened bottle by 13 °C.

Teaching tips

Get members of the class to take the readings and enter them on a pre-prepared table on the blackboard or OHP. The class could prepare suitable graph axes before the experiment and plot the temperatures against time as they are recorded.

Theory

In a greenhouse, visible light passes through the glass (which is, of course, transparent to visible light) and is absorbed by dark coloured surfaces inside. These heat up and re-radiate energy, but at longer wavelengths in the infrared region of the spectrum. This is absorbed by glass and so the greenhouse warms up. The 'greenhouse effect' in the Earth's atmosphere is caused by a number of gases that behave in a similar way to glass, *ie* they are transparent to visible light, but absorb in the infrared. Some of these are listed in the table. It can be seen that carbon dioxide is the most important greenhouse gas because of its relatively high concentration in the atmosphere rather than its intrinsic greenhouse efficiency.

Gas	Relative greenhouse efficiency per molecule	Concentration in the atmosphere / ppm	Relative efficiency x concentration / ppm
Carbon dioxide	1	350	350
Methane	30	1.7	51
Dinitrogen oxide	160	0.31	49.6
Ozone	2 000	0.06	120
CFC 11 (CCl_3F)	21 000	0.000 26	5.46
CFC 12 (CCl_2F_2)	25 000	0.000 24	6

This experiment demonstrates the greenhouse effect caused by the plastic of the bottle. The teacher can explain that gases have the same effect. It also shows the effect of a black surface absorbing and re-radiating energy.

The following articles give useful background for the teacher or post-16 students on the greenhouse effect:

I. Campbell. What on Earth is the greenhouse effect? *Chem. Rev.*, 1991, **1** (2), 2.

I. Campbell. The chemical basis of global warming. *Chem. Rev.*, 1992, **1** (4), 26.

Extensions

Try a thermometer in a glass bottle for comparison with a plastic bottle.

Try sunlight (when available!) instead of the photoflood bulb.

See also demonstration 68.

Further details

This would be an ideal experiment for computer interfacing if thermocouple thermometers were used together with suitable interfacing boxes and software. The graphs could then be plotted on-line on a monitor and hard copies printed for distribution to the class. The book by Robert Edwards, *Interfacing chemistry experiments*. London: RSC, 1993 gives some helpful advice about interfacing.

Safety

Wear eye protection.

The two-holed stoppers are used for the thermometer to prevent pressure build-up inside the bottles caused by the rise in temperature.

It is the responsibility of teachers doing this demonstration to carry out an appropriate risk assessment.

THE ROYAL
SOCIETY OF
CHEMISTRY

68. The greenhouse effect – 2

Topic

Environmental chemistry.

Timing

About 30 min.

Level

Any.

Description

Two beakers, each containing a disk of lead foil, are placed on the bench below a photoflood light bulb. The temperature of the air in each is monitored by a thermometer and eventually becomes steady. Carbon dioxide gas is then led into one of the beakers via a delivery tube and the temperature of this beaker is seen to rise.

Apparatus

▼ One 275 W photoflood light bulb and a suitable plain bulb holder (*ie* one with no shade).

▼ Two identical 250 cm³ beakers.

▼ Two mercury in glass thermometers (0–100 °C).

▼ Two discs of lead foil cut to fit the bases of the beakers.

Chemicals

▼ A source of carbon dioxide gas – a cylinder, chemical generator (such as marble chips and dilute hydrochloric acid in a flask with a delivery tube) or a few chips of dry ice in a flask with a delivery tube.

▼ 1 cm³ of **pentane** and 1 cm³ of **1,1,1-trichloroethane** (optional).

Method

Before the demonstration

Cut two identical discs of lead foil to fit the bases of the beakers. These discs are to absorb and re-radiate radiant energy and they simulate the Earth's surface. Place the beakers side by side on the bench with the lead discs inside and place the photoflood lamp in its holder about 15 cm above. Both beakers should be illuminated equally when the bulb is switched on. Clamp a thermometer with its bulb about 2 cm above the lead disc inside each beaker *(see Fig)*. These thermometers should be chosen so that they read the same temperature in the same surroundings.

Switch on the light and wait until the thermometers have reached steady readings (these will be about 45 °C depending on the distance of the bulb). If these are not the same, move the lamp until they do give identical readings (this is not necessary but makes the results easier to interpret).

The demonstration

Switch on the lamp and take readings from both thermometers every minute for two or three minutes to show that they become steady. Now introduce carbon dioxide

THE ROYAL
SOCIETY OF
CHEMISTRY

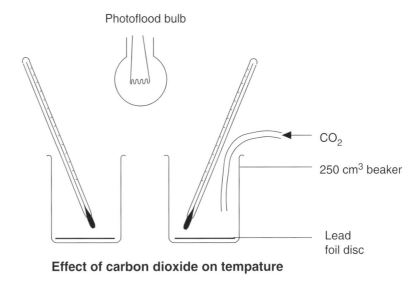

Effect of carbon dioxide on tempature

into one of the beakers using a delivery tube connected to the source, but taking care that the tube does not shade the thermometer from the light. Leave the carbon dioxide flowing slowly to keep the beaker full of carbon dioxide and make up losses by diffusion. A flow rate of about 5 cm^3 per second, estimated by dipping the delivery tube in a beaker of water, is suitable, but this is not critical. The temperature recorded in this beaker will rise and attain a new steady value of about 8 °C above that in the control beaker (which will not change). Stop the flow of carbon dioxide and the temperature will drop again as carbon dioxide diffuses out of the beaker. The whole cycle will take about 15 minutes.

Teaching tips

Get members of the class to take the readings and enter them on a pre-prepared table on the blackboard or OHP. The class could prepare suitable graph axes before the experiment and plot the temperatures against time as they are recorded.

Theory

In a greenhouse, visible light passes through the glass (which is, of course, transparent to visible light) and is absorbed by dark coloured surfaces inside. These heat up and radiate energy but at longer wavelengths in the infrared region of the spectrum. This is absorbed by glass and so the greenhouse warms up. The greenhouse effect in the Earth's atmosphere is caused by a number of gases that behave in a similar way to glass, *ie* they are transparent to visible light but absorb in the infrared. Some of these are listed in the table. It can be seen that carbon dioxide is the most important greenhouse gas because of its relatively high concentration in the atmosphere rather than its intrinsic greenhouse efficiency.

THE ROYAL
SOCIETY OF
CHEMISTRY

Gas	Relative greenhouse efficiency per molecule	Concentration in the atmosphere / ppm	Relative efficiency x concentration / ppm
Carbon dioxide	1	350	350
Methane	30	1.7	51
Dinitrogen oxide	160	0.31	49.6
Ozone	2 000	0.06	120
CFC 11 (CCl_3F)	21 000	0.000 26	5.46
CFC 12 (CCl_2F_2)	25 000	0.000 24	6

The following articles give useful background for the teacher or post-16 students on the greenhouse effect:

I. Campbell. What on Earth is the greenhouse effect? *Chem. Rev.,* 1991, **1** (2), 2.
I. Campbell. The chemical basis of global warming. *Chem. Rev.,* 1992, **1** (4), 26.

Extensions

Other gases can be tried provided that they are denser than air and therefore do not escape too easily from the beaker. Vapours of volatile liquids can also be used, for example 1,1,1-trichloroethane and pentane. About 1 cm^3 of liquid can be added to the beaker using a dropping pipette. The temperature will rise by several degrees as the liquid evaporates and drop again as the vapour diffuses out of the beaker.

Further details

This would be an ideal experiment for computer interfacing if thermocouple thermometers were used together with suitable interfacing boxes and software. The graphs could then be plotted on-line on a monitor and hard copies printed for distribution to the class. The book by Robert Edwards, *Interfacing chemistry experiments.* London: RSC, 1993 gives some helpful advice about interfacing.

Safety

Wear eye protection.
It is the responsibility of teachers doing this demonstration to carry out an appropriate risk assessment.

THE ROYAL
SOCIETY OF
CHEMISTRY

69. The 'breathalyser' reaction

Topic

Reactions of alcohols, everyday chemistry.

Timing

About 5 min.

Level

Post-16 for the reaction of alcohols, any for an illustration of the chemical breathalyser.

Description

A U-tube is packed with orange crystals of potassium dichromate moistened with dilute sulphuric acid. Air saturated with ethanol vapour is blown or sucked over these crystals which turn from brown to green as the alcohol reduces the chromium(VI) to chromium(III).

Apparatus

▼   One U-tube (length of arm about 10 cm). Each arm should have a one-hole bung fitted with a short length of glass tube.

▼   Two 250 cm³ conical flasks each with a two-holed bung fitted with one long and one short length of glass tube.

▼   One rubber bung with a single hole fitted with a short length of glass tube.

▼   One plastic sandwich bag.

▼   Rubber tubing to make connections.

▼   Access to a filter pump.

▼   Cable tie or length of thread.

Chemicals

The quantities given are for one demonstration.

▼   About 100 cm³ of **ethanol.**

▼   About 30 g of **potassium dichromate** (potassium dichromate(VI), $K_2Cr_2O_7$) crystals.

▼   About 3 cm³ of 2 mol dm⁻³ **sulphuric acid**.

▼   Bottles of ethanal (acetaldehyde) and of an approximately 2 mol dm⁻³ solution of ethanoic acid (acetic acid) (optional).

Method

Before the demonstration

Wearing plastic gloves, weigh out into a beaker sufficient potassium dichromate crystals to half fill the U-tube (about 30 g). Add 2 mol dm⁻³ sulphuric acid to the crystals, in the ratio 1 cm³ of acid to 10 g of dichromate and mix thoroughly. This should produce moistened crystals of potassium dichromate. Put these into the U-tube, tapping the tube gently to pack them down.

     Place ethanol in the conical flask to such a depth that the longer of the glass tubes

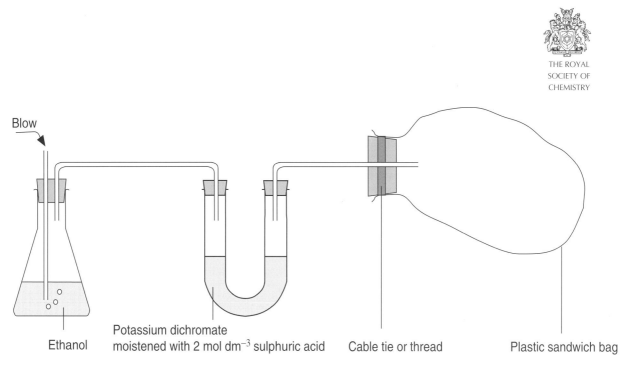

Blow

Ethanol

Potassium dichromate
moistened with 2 mol dm^{-3} sulphuric acid

Cable tie or thread

Plastic sandwich bag

Fig. 1 The breathalyser

Cable tie

Fig. 2 The cable tie

is below the surface of the ethanol and the shorter one is not.
Attach a plastic sandwich bag to a one-holed rubber bung by gathering the neck of
the bag around the bung and attaching it with a cable tie or by winding thread
around it *(see Fig. 2)*.

The demonstration

Blow air by mouth through the ethanol in the flask so that the air in the flask is
saturated with ethanol vapour. Make sure the plastic bag is deflated. Connect the
conical flask, U-tube and plastic bag as shown in *Fig. 1*. Blow into the ethanol-
containing flask so that the breath bubbles through the ethanol, passes over the
acidified potassium dichromate and into the plastic bag. The dichromate crystals in
the arm of the U-tube nearest the ethanol will turn brown. This is caused by a
mixture of unreacted orange crystals and green chromium(III), a product of the
reaction.

   If desired, continue the reaction by removing the plastic bag and connecting the
apparatus to a filter pump to draw more ethanol vapour over the crystals. The brown
colour will spread and eventually turn green although this will take several minutes.
Remove the stoppers from the U-tube and pass it round the class for the students to
smell the products of the reaction. Let them compare the smell with the smells of
ethanal and ethanoic acid solution- possible products of the reaction.

Visual tips

A white background will make the colour changes easier to see and a bottle of
unreacted potassium dichromate is useful for comparison.

THE ROYAL
SOCIETY OF
CHEMISTRY

Teaching tips

The teacher could also demonstrate the reaction of ethanol with acidified potassium dichromate solution in a test-tube in the usual way if the students are not familiar with it.

This could be a suitable opportunity to discuss the dangers of drinking and driving.

Balancing the equations can be set as homework for post-16 students.

Theory

The reactions are:

$$Cr_2O_7^{2-}(s) + 3C_2H_5OH(g) + 8H^+(aq) \rightarrow 3CH_3CHO(g) + 2Cr^{3+}(aq) + 7H_2O(l)$$

producing ethanal, followed by:

$$Cr_2O_7^{2-}(s) + 3CH_3CHO(g) + 8H^+(aq) \rightarrow 3CH_3CO_2H(l) + 2Cr^{3+}(aq) + 4H_2O(l)$$

producing ethanoic acid. Or, overall:

$$2Cr_2O_7^{2-}(s) + 3C_2H_5OH(g) + 16H^+(aq) \rightarrow 3CH_3CO_2H(l) + 4Cr^{3+}(aq) + 11H_2O(l)$$

Further details

Chemical breathalysers are no longer in use; roadside screening is done using a tester based on a fuel cell while testing at the police station for use in evidence is done by measuring infrared absorption. Further confirmation, if requested by the suspect, may be done by GLC. Details can be found in C. B. Faust, *Modern chemical techniques* London: RSC, 1992.

It is illegal to drive with a blood alcohol concentration of more than 80 mg per 100 cm³. An equilibrium is established in the lungs between alcohol dissolved in blood and alcohol in the breath (an interesting example of Henry's law!). The equivalent breath concentration is 35 microgrammes of alcohol per 100 cm³ of breath.

Safety

Wear eye protection.

Some teachers may wish to include an extra conical flask in the apparatus to act as a trap to guard against the possibility of suckback while blowing into the ethanol (*Fig. 3*).

It is the responsibility of teachers doing this demonstration to carry out an appropriate risk assessment.

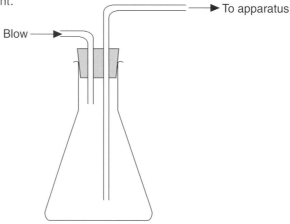

Fig. 3 Preventing suckback flask

THE ROYAL
SOCIETY OF
CHEMISTRY

70. The electrolysis of water – exploding bubbles of oxygen and hydrogen

Topic

Electrolysis, combustion reactions.

Timing

About 10 min.

Level

Pre-16.

Description

A dilute solution of sodium sulphate containing a little universal indicator is electrolysed using platinum electrodes. The pH changes at the electrodes can be seen. The hydrogen and oxygen evolved at the electrodes are mixed and used to blow soap bubbles. These bubbles can be exploded giving a loud 'crack'.

Apparatus

▼ Variable DC power pack capable of supplying a current of at least 1 A at 12 V.

▼ Connecting leads and crocodile clips.

▼ Ammeter (0–1 A) (optional).

▼ Two pieces of platinum wire each about 10 cm long. Only one piece of platinum is essential, the other can be replaced with iron, copper or nichrome wire.

▼ One clear glass jar (about 400 cm^3) as used to store powders.

▼ One one-holed rubber bung to fit the jar.

▼ One short length of glass tube.

▼ One length of flexible plastic tubing.

▼ One 250 cm^3 beaker.

▼ Bunsen burner.

▼ Spatula with a spoon-type end or a teaspoon.

Chemicals

The quantities given are for one demonstration.

▼ About 10 g of sodium sulphate (Na_2SO_4).

▼ A little washing up liquid.

▼ A little universal indicator solution.

Method

Before the demonstration

Set up the apparatus shown in the *figure*. The platinum wires can be inserted through the rubber bung by making holes with a piece of stiff wire that has been heated to red heat in a blue Bunsen burner flame. Check that the apparatus is gas tight and seal the wires in with Vaseline or Blu-tac if necessary.

Make a solution of about 10 g of sodium sulphate in 500 cm³ of water.

Light a Bunsen burner well away from the electrolysis apparatus!

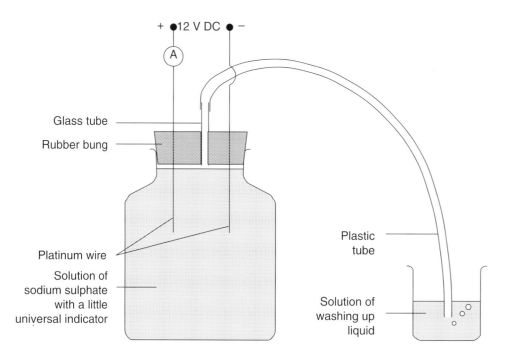

Splitting water to produce hydrogen and oxygen

The demonstration

Fill the jar with sodium sulphate solution leaving only enough air space to fit the bung. Add a few drops of universal indicator solution to the solution so that the green colour is clearly visible. Fit the bung and connect the electrodes to a power pack set at 12 V DC. Connect an ammeter in series in the circuit if desired. Switch on the power pack; the current should be about 1 A. The indicator will turn blue around the cathode due to the formation of OH⁻(aq) ions and yellow around the anode due to the formation of H⁺ (aq) ions. Bubbles of oxygen will form at the anode and **hydrogen** at the cathode. Point out that there is about twice as much hydrogen as oxygen.

Observe these changes for a couple of minutes to allow air in the delivery tube to be displaced by the mixture of hydrogen and oxygen. Now place the end of the delivery tube in a beaker of water containing a little washing up liquid. Bubbles will form and collect at the surface of the water. Scoop up some bubbles in a spatula or spoon and hold them in the Bunsen flame. They will explode with a sharp crack, which is impressive considering the small amount of gas mixture. If the bubbles do not explode, wait a little longer for the gas mixture to displace the air from the delivery tube. Do not attempt to ignite bubbles at the end of the delivery tube.

Visual tips

A white background is essential if the colour changes are to be seen.

Teaching tips

Point out that electrical energy is required to split the water, that water is reformed in the explosion and that energy is given out. Producing hydrogen by electrolysis of

THE ROYAL
SOCIETY OF
CHEMISTRY

water for use as a fuel is one suggestion for an alternative fuel economy to that based on petrochemicals (provided that the electricity has not been generated by burning oil or gas!).

Theory

The electrode reactions are as follows:

at the anode:

$$2H_2O(l) \rightarrow O_2(g) + 4H^+(aq) + 4e^-$$

at the cathode:

$$4H_2O(l) + 4e^- \rightarrow 2H_2(g) + 4OH^-(aq)$$

Overall:

$$2H_2O(l) \rightarrow O_2(g) + 2H_2(g)$$

Extensions

The gases could be collected in a separate experiment, using a Hoffman voltameter or an electrolysis cell, to confirm that the volume of hydrogen is double that of oxygen.

Further details

The anode must be made of platinum wire – other electrodes tend to react with the oxygen that is produced. Copper, iron or nichrome work satisfactorily as the cathode and can be used if only one piece of platinum is available. The apparatus could be modified to use platinum electrodes from a hydrogen electrode for example.

Safety

Wear eye protection.
   It is the responsibility of teachers doing this demonstration to carry out an appropriate risk assessment.

THE ROYAL
SOCIETY OF
CHEMISTRY

71. The preparation of nitrogen monoxide and its reaction with oxygen

Topic

Oxides of nitrogen, volumes of reacting gases, equilibria, environmental chemistry.

Timing

About 10 min.

Level

Pre-16 for the preparation of nitrogen monoxide, post-16 for the reacting volumes and equilibrium.

Description

Nitrogen monoxide (nitrogen(II) oxide or nitric oxide) is prepared by the reaction of iron(II) sulphate and sodium nitrate(III) in acid solution. The reaction with oxygen to give brown nitrogen dioxide is demonstrated either by allowing it to come into contact with air or by using measured volumes of nitrogen monoxide and oxygen in gas syringes. The expected volume of nitrogen dioxide is not obtained because of the equilibrium between nitrogen dioxide and dinitrogen tetroxide. The resulting mixture can be dissolved in water leaving a very small residual volume of gas. This can be used as an illustration of Gay-Lussac's law. An equilibrium constant for the dimerisation of nitrogen dioxide can be estimated.

Apparatus

▼ Two 100 cm³ glass gas syringes. Smaller ones can be used.

▼ Two three-way glass stopcocks (optional).

▼ One 250 cm³ conical flask with a two-holed stopper.

▼ Tap funnel.

▼ A length of glass tubing with a right angled bend.

▼ Rubber tubing to make connections between the delivery tube, stopcocks and gas syringes.

▼ Glass trough or washing up bowl.

▼ Two test-tubes with stoppers.

▼ One 20 cm³ plastic syringe with a hypodermic needle.

▼ Access to an overhead projector (optional).

▼ Access to a fume cupboard.

Chemicals

The quantities given are for one demonstration.

▼ About 150 g of iron(II) sulphate-7-water (hydrated ferrous sulphate, $FeSO_4.7H_2O$).

▼ About 50 g of **sodium nitrate(III)** (sodium nitrite, $NaNO_2$)

▼ About 75 cm³ of **concentrated hydrochloric acid**.

THE ROYAL
SOCIETY OF
CHEMISTRY

▼   About 75 cm³ of deionised water.

▼   Access to a source of oxygen – either a cylinder or a chemical generator (for example, drip hydrogen peroxide onto manganese dioxide using apparatus similar to the one used to prepare the nitrogen monoxide).

NB Oxygen from either source can be stored conveniently in a 'gas bag'. See demonstration 19.

Method

Before the demonstration

Add 75 cm³ of concentrated hydrochloric acid to 75 cm³ of deionised water to make 150 cm³ of approximately 5.5 mol dm⁻³ hydrochloric acid.

Fill a 'gas bag' with oxygen from a cylinder or chemical generator. It will be necessary to partly fill and flush out the bag three or four times to remove air from the system. This is easier if a three-way stopcock is fitted to the bag but this is not essential.

Set up the gas generator as in *Fig. 1*. Place about 150 g of iron(II) sulphate in the flask and add about 150 cm³ of the 5.5 mol dm⁻³ hydrochloric acid. This will give a saturated solution of iron(II) sulphate.

Make a solution of 50 g of sodium nitrate(III) in 100 cm³ of water and place this in the tap funnel.

Fill the plastic syringe with water to above the 20 cm³ mark. Fit the hypodermic needle and expel any air and excess water until there is exactly 20 cm³ of water in the syringe.

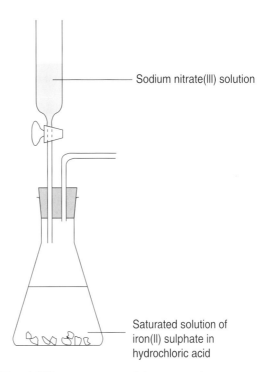

Sodium nitrate(III) solution

Saturated solution of iron(II) sulphate in hydrochloric acid

Fig. 1 Nitrogen monoxide generator

The demonstration

Working in a fume cupboard, run a little sodium nitrate(III) solution from the tap funnel into the solution of iron(II) sulphate in the flask. Bubbles of nitrogen monoxide are produced and the green solution turns brown. The gas is colourless, but reacts

THE ROYAL
SOCIETY OF
CHEMISTRY

immediately with oxygen in the air to form brown nitrogen dioxide. Wait for a few minutes until the brown gas has been displaced and the flask is filled with colourless nitrogen monoxide.

Collect two test-tubes of the nitrogen monoxide by displacement of water. Stopper the tubes while their mouths are still under water. Show that the gas is colourless and then remove the stopper from the first tube to allow air in. The nitrogen monoxide will react immediately with oxygen in the air to produce brown **nitrogen dioxide**. Unstopper the second tube of gas while the mouth of the tube is still under water. Leave the tube clamped vertically with its mouth under water. The water does not rise appreciably up the tube, showing that nitrogen monoxide is insoluble. Still keeping the mouth of the tube under water, manoeuvre a piece of full range indicator paper into the tube. This will not change colour, showing that the gas is neutral.

Fill one of the gas syringes with nitrogen monoxide. It will be necessary to part-fill and flush out the syringe a few times to ensure that there is no air left in the system. A three-way stopcock is helpful but not necessary. When the air has all been displaced, the gas in the syringe will be colourless. Fill to about the 90 cm^3 mark.

Fill the second syringe with oxygen either from the gasbag or from a cylinder or generator. Part fill and flush three or four times to remove air as described above. A three-way stopcock is helpful but not necessary. Fill the syringe to the 40 cm^3 mark.

Expel a little of the gas from the syringe of nitrogen monoxide so that it is filled to the 80 cm^3 mark. Join the two syringes with a short length of rubber tubing and push a little of one of the gases into the other *(Fig. 2)*. The two gases will react immediately to give brown nitrogen dioxide and the volume will shrink. Manipulate the syringe plungers to pass the gas from syringe to syringe two or three times to ensure complete reaction and note the final volume, which will be about 45 cm^3. Now use the 20 cm^3 syringe and hypodermic needle to inject 20 cm^3 of water into the connected gas syringes by sticking the needle through the rubber tubing. The gas volume will shrink and the brown colour will disappear as the nitrogen dioxide and dinitrogen tetroxide dissolve in the water. Manipulate the syringe plungers to pass the water and gas from syringe to syringe two or three times to ensure complete dissolution. Note the final volume, which will be about 27 cm^3 including the 20 cm^3 of water. Squirt a little of the resulting solution onto indicator paper and note the strongly acidic pH.

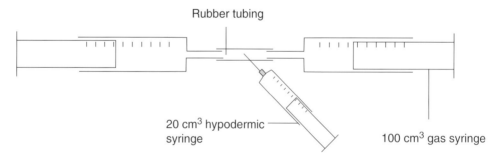

Rubber tubing

20 cm^3 hypodermic syringe

100 cm^3 gas syringe

Fig. 2 Formation and dissolution of nitrogen dioxide

Visual tips

A white background is best for viewing the colour change.

The part of the experiment involving the gas syringes can be done on an overhead projector in which case smaller syringes may be more convenient.

Teaching tips

Teachers may prefer to omit the quantitative part of the demonstration and simply show the properties of nitrogen monoxide and its reaction with oxygen in the air.

When reading the volumes of gas in the syringes, stress that the pressure has

THE ROYAL
SOCIETY OF
CHEMISTRY

remained constant (atmospheric) and twist the barrels of the syringes to ensure that they do not stick.

Theory

The reaction to prepare the nitrogen monoxide is:

$$6FeSO_4(aq) + 12NaNO_2(aq) \rightarrow 2Fe_2O_3(s) + Fe_2O_3.N_2O_5(aq) + 10NO(g) + 6Na_2SO_4(aq)$$

The reaction of nitrogen monoxide with oxygen is:

$$2NO(g) + O_2(g) \rightarrow 2NO_2(g)$$

So one might expect 80 cm^3 of nitrogen monoxide to react with 40 cm^3 of oxygen to give 80 cm^3 of nitrogen dioxide.
    But nitrogen dioxide exists in equilibrium with its dimer, dinitrogen tetroxide:

$$2NO_2(g) \rightleftharpoons N_2O_4(g)$$

and this reduces the measured volume.
    Nitrogen dioxide reacts with water to give a mixture of nitric(V) and nitric(II) acids:

$$2NO_2(g) + H_2O(l) \rightarrow HNO_3(aq) + HNO_2(aq)$$

Extensions

It is possible to use the measured final volume of gas to estimate the equilibrium constant for the reaction

$$2NO_2(g) \rightleftharpoons N_2O_4(g)$$

as follows. The resulting figure roughly agrees with the literature value of 8.7 atm^{-1} at 298 K.

$$2NO_2(g) \rightleftharpoons N_2O_4(g)$$

Start	80 cm^3	0
At eqm	80 $-2x$	x

Thus the total volume of the gas mixture is $80 - 2x + x = 45$

$x = 35$

So at equilibrium, the volume of N_2O_4 is 35 cm^3 and of NO_2 is 10 cm^3.

Hence the partial pressures are:

N_2O_4 : 35/45 atm
NO_2 : 10/45 atm

$$K_p = \frac{p_{N_2O_4}}{p^2_{NO_2}}$$

$K_p = 15.8$ atm^{-1}

Further details

Nitrogen monoxide is formed in lightning flashes and is eventually oxidised to nitric acid in the atmosphere. It is also formed at high temperatures in internal combustion engines. The resulting nitrogen dioxide is a significant component of photochemical smog.

An alternative method of preparing nitrogen monoxide is by the reaction of copper turnings with 50 % nitric acid, but this gas is less pure than that produced by the method given above.

The residual gas left after absorbing the equilibrium mixture in water may contain air due to incomplete flushing out of the apparatus during filling. It may contain unreacted oxygen or nitrogen monoxide due to measuring errors or due to the volume of gas contained in the syringe nozzles. It may contain nitrogen monoxide from the disproportionation of nitric(III) acid

$$3HNO_2(aq) \rightarrow HNO_3(aq) + 2NO(g) + H_2O(l)$$

Safety

Wear eye protection.

Empty the syringes in the fume cupboard.

It is the responsibility of teachers doing this demonstration to carry out an appropriate risk assessment.

THE ROYAL
SOCIETY OF
CHEMISTRY

72. Reactions of the alkali metals

Topic

Group I metals (alkali metals), the Periodic Table.

Timing

About 1 h.

Level

Pre-16.

Description

The reactions of lithium, sodium and potassium with air, water and chlorine are demonstrated along with some of the physical properties of the metals: density, softness and melting temperatures.

Apparatus

▼ Three gas jars with lids.

▼ Three glass troughs of capacity about 5 dm³. (It is possible to manage with just one.)

▼ Sheet of glass or perspex to act as a lid to cover the troughs.

▼ Filter paper.

▼ Three tin lids – for example the lids from old sweet tins. Ensure that any plastic seals have been removed from inside the rims and that any paint has also been removed. This can be done by heating strongly in a fume cupboard. Clean the lids with emery paper.

▼ Three petri dishes with lids.

▼ Scalpel or other sharp knife.

▼ Tweezers.

▼ Three deflagrating spoons.

▼ Circuit board (or similar) with batteries, leads and bulb to show electrical conduction.

▼ Disposable plastic gloves.

▼ Three boiling tubes.

▼ Wooden spills.

▼ Bunsen burner.

▼ Heat-proof mat.

▼ Safety screen.

If the chlorine is to be generated chemically:

▼ One 250 cm³ conical flask with a two holed stopper to fit.

▼ A tap funnel.

▼ One piece of glass tubing with a right angled bend.

▼ Access to fume cupboard.

Chemicals

The quantities given are for one demonstration.

▼   One piece of lithium, sodium and potassium, each about 1 cm³ in volume. These are stored under paraffin oil.

▼   Access to a source of **chlorine** – either a cylinder or a chemical generator (use a tap funnel to drip **concentrated hydrochloric acid** onto solid potassium manganate(VII) (**potassium permanganate**)). If chlorine generated chemically is to be used, about 10 g of potassium manganate(VII) (potassium permanganate, $KMnO_4$) and 100 cm³ of concentrated hydrochloric acid will be needed.

▼   Universal indicator solution and paper with appropriate colour charts.

▼   A few cm³ of hexane or other solvent for the paraffin oil.

Method

Before the demonstration

Working in a fume cupboard, fill the three gas jars with chlorine by upward displacement of air and close each one with a lightly greased lid. The contents of the jars should be noticeably green. Well sealed jars will keep for a few hours.

Set up a series circuit consisting of a bulb, suitable number of batteries and two flying leads to be used to test electrical conductivity. A circuit board is a convenient way to do this.

The demonstration

This is in four parts: physical properties; heating in air; reaction with chlorine; and reaction with water. It is suggested that each of these is done for all three metals in the above order so that the metals can be compared easily in each category and so that the demonstration builds up to a climax. Within each part, do the metals in the order lithium, sodium, potassium to stress the Periodic Table order and also to leave the most reactive until last.

1. Physical properties

Do the following for each metal:

Using tweezers remove a piece about 5 mm x 5 mm x 5 mm from the paraffin oil (explaining why it is stored in this way and pointing out that lithium floats on the oil). Do not bother to remove the oil at this stage as a thin film will continue to protect the metal. Wearing plastic gloves, cut off a small piece (which can be used later) to demonstrate the ease with which the metals can be cut. Point out the differences in ease of cutting between the metals themselves and compare them with a more typical metal such as iron. All three alkali metals can be cut with a sharp knife, lithium being the hardest and potassium being the softest. At this point is also possible to show that potassium can be moulded easily by squeezing between gloved fingers, sodium is harder to mould and lithium can barely be moulded at all. If desired, place a piece of each metal in a petri dish and pass it round the class for the students to look at closely and to feel its lightness. Warn the class not to touch the metals and tape the lids closed if you have any doubts about their reliability.

Drop each metal from a height of a few cm onto the bench; the gentle impacts indicate their low density. Compare with a similar sized piece of a typical metal such as iron. If desired, each metal could be weighed to give a very rough estimate of density, again comparing with a similar sized lump of, say, iron.

Use the circuit board to show that each metal conducts electricity well.

THE ROYAL
SOCIETY OF
CHEMISTRY

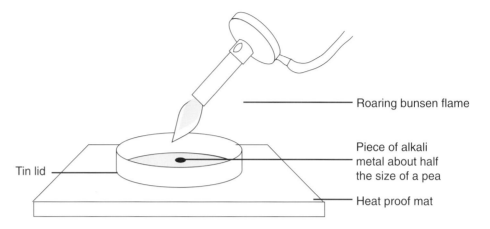

Fig. 1. Heating the metal

2. Heating in air

Do the following for each metal:

Cut a piece about 2 mm x 2 mm. Show how shiny the freshly cut face is and how rapidly it tarnishes. It may be necessary to do this two or three times to small groups of the class as it will be difficult to see from a distance. Using tweezers, rinse the piece in hexane to remove the oil and wave it around in the air to dry. Place it on an inverted tin lid standing on a heatproof mat and heat from above with a roaring Bunsen flame using the part of the flame about 1 cm beyond the tip of the blue cone (*Fig. 1*). Point out that the metal melts into a ball and begins to burn. Potassium will melt most rapidly and lithium will take longest to melt. Remove the Bunsen burner and allow the metal to burn away and note the colours of the flames – red for lithium, orange for sodium and lilac for potassium. Note the greyish colour of the residue and, when it is cool, test the residue by touching it with moist indicator paper. It will be very alkaline. There will not appear to be a great deal of difference in reactivity between the metals in this reaction.

3. Reaction with chlorine

This can be done in a variety of ways. The simplest is to place a small piece of the metal on a deflagrating spoon. Heat it in a Bunsen flame until the metal begins to burn and plunge the spoon into a gas jar of chlorine. The metal will combine vigorously with chlorine, producing clouds of white metal chloride 'smoke'. Point out that this smoke is not a gas but is a cloud of solid particles which soon settle to the bottom of the gas jar. Unfortunately, most deflagrating spoons are made of iron and a brown smoke of iron(III) chloride is also produced.

One can avoid the production of iron(III) chloride by using a glass deflagrating spoon (if available) or by improvising one by cutting off the bottom 2 mm or so off a small test-tube and holding this with tweezers.

A second alternative is to place a small piece of the metal (free of oil) in the base of the gas jar of chlorine (see *Fig. 2*) on a piece of ceramic material (to protect the glass). Use a dropping pipette to drop one or two drops of water onto the metal so that it reacts exothermically with the water. The heat generated will be enough to start the reaction between the metal and the chlorine in the case of sodium and of potassium. Clouds of white chloride will be produced. Lithium will not react with chlorine using this method while potassium may start to react spontaneously without the need for the water.

Whichever method is used, it will be possible to see that the order of reactivity is potassium > sodium > lithium.

(4) Reaction with water

Fill the trough about half full of water. Place a piece of metal (free of oil) about the size of a rice grain on the water surface. Place the glass lid on the trough as the metals often 'spit'. Lithium will float and fizz quietly producing hydrogen. Sodium will float and fizz rapidly, moving around on the surface. Potassium will float, fizz , move around and set on fire, burning with a lilac flame. This experiment gives the clearest indication of the order of reactivity of the metals. After each reaction, test the water by adding a little universal indicator solution. The water will be alkaline.

Repeat the reaction using pieces of metal about half the size of a rice grain in a boiling tube half full of water. Point the tube away from yourself and the audience in case of spitting. After a few seconds, test the gas with a lighted spill. A loud pop will be heard showing that it is hydrogen. The water in the tube will get hot, noticeably so in the case of potassium, and a member of the audience can be asked to feel the tube, after the reaction has stopped, to confirm the temperature rise.

Alternatively, float a piece of filter paper on the water surface so that it is soaked with water. Place the metal on the filter paper. In this case, both potassium and sodium will catch fire, but lithium will not. The metals may float away from the paper, but both sodium and potassium will continue to burn. This is because there is less water in which to dissipate the heat of the reaction so the sodium reaction becomes hot enough to ignite the hydrogen. The characteristic colours of the sodium and potassium flames look spectacular in a darkened room.

Visual tips

The reaction with water can be demonstrated on an OHP. Place the trough on the OHP and focus on a matchstick floating on the water. Remove the matchstick and add a few drops of suitable indicator (universal indicator, or another indicator familiar to the students). If necessary, add a few drops of dilute acid so that the indicator is in its acid form. Drop in the metal and the fizzing will be seen, as well as trails of colour which form behind the metals as they float around. These trails will be in the alkaline colour of the indicator.

Teaching tips

Younger students might enjoy writing a mock newspaper account of the consequences of reacting francium with water (if it was possible to obtain enough of it).

In each case, there is sufficient information to identify the products of the reaction and it can be a useful exercise to build up word and then balanced symbol equations with the students.

It is a good idea if students have ready a prepared table in which to fill in their observations as they watch the demonstration.

A data book exercise could be set in which the students look up and tabulate data such as density and melting point to support the estimates made in the demonstration. They could plot graphs of the data to show up trends.

Theory

The reactions are as follows (using M to represent any of the alkali metals).

With oxygen:

$$4M(s) + O_2(g) \rightarrow 2M_2O(s)$$

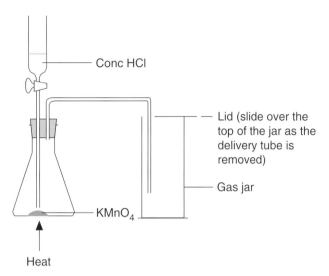

Fig. 2. Making Chlorine

In the case of sodium, some peroxide (Na_2O_2) will also be formed and in the case of potassium and other Group I metals, superoxides (MO_2) will be formed too. Superoxides are unusual for compounds of Group I elements because they are coloured. They are more likely to form with the larger Group I metal ions since there is more room for the large superoxide ion in the crystal lattice.

With chlorine:

$$2M(s) + Cl_2(g) \rightarrow 2MCl(s)$$

With water:

$$2M(s) + 2H_2O(l) \rightarrow 2M(OH)(aq) + H_2(g)$$

The usual explanation given for the reactivity of Group I metals increasing on going down the Group is that the outer electron is further away from the nucleus and is therefore more easily removed (*ie* that the first ionisation energy gets lower). This is true, but it is only one of many factors involved in determining ΔH for the reactions. It should also be remembered that what is being observed in these reactions is often related to the rate of the reaction while the ionisation energy is related to ΔH. These two parameters are not linked except that a strongly exothermic reaction, once started, will give out heat which will then speed up the reaction. The clearly observed relationship between position in the Group and reactivity is therefore partly fortuitous.

Extensions

Videos are available that show the reactions of alkali metals including those of rubidium and caesium with water, for example *Elements organised* from the Open University. One of these could be shown after the demonstration to reinforce the reactivity trend. The reaction of caesium with water is violent enough to shatter the trough.

Further details

One teacher reports that adding a few drops of washing up liquid to the water prevents the metal piece sticking to the side of the trough when the reactions with water are being shown.

For more details about the reactions of alkali metals and a discussion of safety issues, see the article entitled 'The exploding metals' by Gareth D John in *Sch. Sci. Rev.*, 1980, **62** (219), 279.

Safety

Wear eye protection.

The alkali metals are flammable solids that react with water to produce alkaline solutions and hydrogen. Excess metals should be disposed of by reacting with propan-2-ol (isopropanol).

It is the responsibility of teachers doing this demonstration to carry out an appropriate risk assessment.

THE ROYAL
SOCIETY OF
CHEMISTRY

73. Sulphur

Topic

Properties of sulphur, allotropy, the relationship between properties and structure.

Timing

About 20 min.

Level

Pre-16.

Description

Sulphur is heated gently and the various forms are:

▼　　an amber, mobile liquid consisting of S_8 rings;

▼　　a viscous liquid consisting of long tangled chains of sulphur; and

▼　　a dark, mobile liquid consisting of shorter sulphur chains.

The first form is allowed to cool and needle-shaped crystals of monoclinic sulphur form.
　　The third form is plunged into cold water and plastic sulphur is formed.
　　Sulphur is dissolved in toluene or xylene and the solvent allowed to evaporate. Crystals of rhombic sulphur form.

Apparatus

▼　　Two boiling tubes.

▼　　Holder to hold the tubes while heating them.

▼　　One 250 cm³ conical flask with cork to fit.

▼　　One 250 cm³ beaker.

▼　　One 1 dm³ beaker.

▼　　Petri dish or watch glass.

▼　　Microscope – ideally a projection or video microscope.

▼　　Bunsen burner, tripod and gauze.

▼　　Heat-proof mat.

▼　　Access to a fume cupboard.

Chemicals

The quantities given are for one demonstration.

▼　　About 60 g of powdered roll **sulphur**. Note that 'flowers of sulphur' is not suitable because it contains a lot of insoluble amorphous sulphur.

▼　　About 700 cm³ of cooking oil or other high boiling point oil.

▼　　About 100 cm³ of **toluene** (methylbenzene) or **xylene** (dimethylbenzene).

▼　　Filter paper – about 18 cm in diameter.

THE ROYAL
SOCIETY OF
CHEMISTRY

Method

Before the demonstration

Two thirds fill a 1 dm^3 beaker with cooking oil and heat to about 130 °C. Half fill a 250 cm^3 beaker with cold water.

The demonstration

Two-thirds fill two boiling tubes with broken/powdered roll sulphur (about 20 g in each tube) and place in the oil bath. The sulphur will melt to a clear, amber, transparent, mobile liquid. This will take about 15 min. Some teachers may wish to pre-prepare at least one tube to save time. Remove one boiling tube and pour the sulphur into a filter paper cone held together by a paper clip (*Fig. 1*) and supported in a beaker. Allow the sulphur to cool and solidify. Break the crust with a spatula and pour off any remaining liquid. Needle-shaped crystals of monoclinic sulphur will be seen. When cool, the cone can be passed around the class.

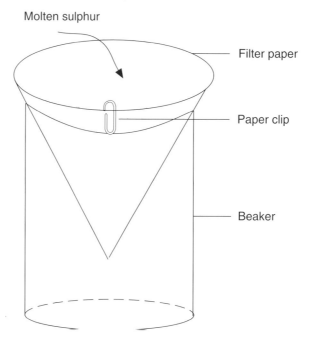

Fig. 1 Preparation of monoclinic sulphur

Take the second tube and, using a test-tube holder, heat it gently over a small Bunsen flame, keeping the contents of the tube moving to prevent local overheating. The liquid gets darker and, fairly suddenly, becomes a viscous gel-like substance. This occurs at about 200 °C. The tube can be inverted and the sulphur will remain in it. Show that the mobile liquid re-forms on cooling.

Further heating beyond the gel-like stage liquefies the sulphur again to a very dark red-brown liquid (the colour of bromine). Note that during heating it is probable that the sulphur will catch fire and **sulphur dioxide** will be produced. Have a heat-proof mat to hand to place over the mouth of the tube to extinguish the blue flames.

When the sulphur begins to boil, pour the liquid sulphur in a slow stream into a beaker of cold water. A mass of brown plastic sulphur will form. Allow this to cool thoroughly, taking care because the inside of the plastic sulphur may remain molten after the outside has solidified. Remove the plastic sulphur from the water and show that it is rubbery – it can be stretched and will return to its original shape. After about

THE ROYAL
SOCIETY OF
CHEMISTRY

half an hour it will be noticeable that the shiny surface of the plastic sulphur is beginning to dull and some of the elasticity is lost as it begins to turn back to the more stable rhombic sulphur. Leave until the following lesson to monitor the progress of this change. This will be very noticeable after a week or so but complete change will take a long time.

Working in a fume cupboard, put about 10 g of powdered roll sulphur into a conical flask and add about 100 cm³ of toluene or xylene. Leave the sulphur to dissolve. This will take some several minutes; warming to about 50 °C will speed up dissolution. Some teachers may wish to prepare the solution before the demonstration to save time. Pour a little of the solution into a petri dish, watch glass or microscope slide and leave it in the fume cupboard for the solvent to evaporate. This will take about 10 min. Small crystals of rhombic sulphur will form. These can be viewed under a microscope. The class will need to file past and view them in turn. It is worth the teacher selecting a well-formed crystal for viewing.

Visual tips

A projection microscope or video microscope can be used to show the shape of the rhombic crystals to the whole class.

Teaching tips

Some stages of this demonstration are time-consuming, *eg* melting the sulphur in the oil bath, dissolving the sulphur in the toluene or xylene and evaporating the solvent. Some teachers may prefer to melt some sulphur before the lesson and to prepare rhombic crystals before the lesson to save time. In the latter case, slower evaporation (which can be brought about by covering the petri dish with filter paper with a few holes in) will produce larger crystals. Particularly large and/or well-formed crystals could be retained as examples for future use.

Theory

Powdered sulphur consists of puckered S_8 rings in the shape of crowns. These can be packed together in two different ways – to form rhombic crystals and to form needle-shaped monoclinic crystals *(Fig. 2)*. Below about 96 °C, rhombic sulphur is the more stable allotrope. On melting at about 118 °C, sulphur first forms a mobile, amber liquid containing S_8 rings. If this is allowed to cool, monoclinic sulphur forms as crystallisation occurs above 96 °C. Monoclinic sulphur will turn slowly into the more stable rhombic form.

Further heating of the S_8-containing liquid breaks the rings into S_8 chains which may join to form longer chains which tangle, causing an increase in viscosity. Further

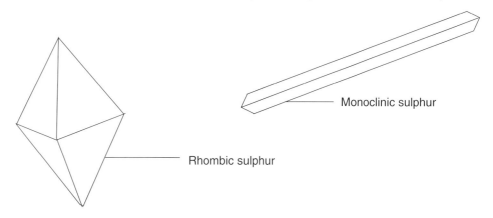

Monoclinic sulphur

Rhombic sulphur

Fig. 2 Different crystal forms of sulphur

heating breaks these chains into shorter ones, perhaps as short as S_2, and the viscosity decreases again. Rapid cooling of this liquid traps the resulting solid sulphur in the tangled chain state – this is plastic sulphur. On stretching, the chains uncoil and on releasing the tension they return to the partly coiled state (*see scheme*).

If solid sulphur is formed below 96 °C, by evaporating the solvent from a solution for example, the stable rhombic form is produced (*see Fig. 3*).

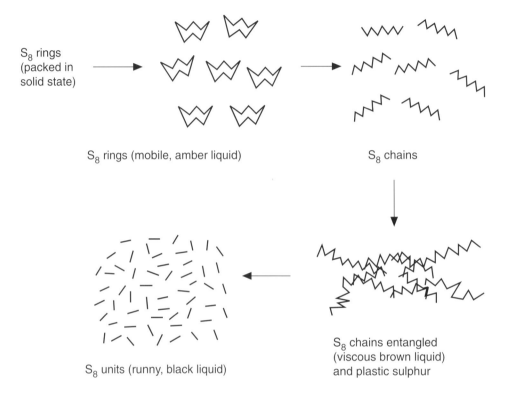

S_8 rings (packed in solid state)

S_8 rings (mobile, amber liquid)

S_8 chains

S_8 units (runny, black liquid)

S_8 chains entangled (viscous brown liquid) and plastic sulphur

Effect of heat on sulphur

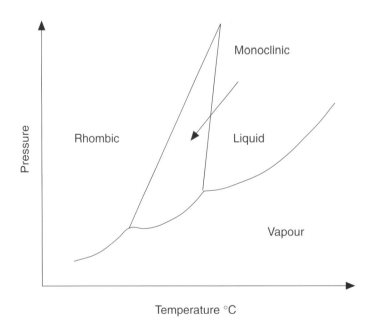

Fig. 3 Phase diagram for sulphur

THE ROYAL
SOCIETY OF
CHEMISTRY

Extensions

To avoid the risk of the sulphur catching fire during heating, a mineral wool plug can be loosely fitted in the mouth of the boiling tube.

Further details

Carbon disulphide is a better solvent for making rhombic sulphur, but its smell, toxicity and flammability make it unsuitable for use in schools.

Very slow heating is essential if all of the changes on heating sulphur are to be seen clearly. Sulphur is a poor thermal conductor, hence the changes can overlap one another if the heating is too fast. It is difficult to heat slowly enough using a Bunsen burner.

Monoclinic crystals can be formed by allowing a hot solution of sulphur in boiling xylene to cool so that crystallisation starts at above 96 °C. Full details can be found in, for example, *Revised nuffield chemistry teachers' guide II*, p138. London: Longman, 1978.

Safety

Wear eye protection.

It is the responsibility of teachers doing this demonstration to carry out an appropriate risk assessment.

74. The thermit reaction

Topic

Reactions of metals, displacement reactions.

Timing

About 5 min.

Level

Pre-16.

Description

A stoichiometric mixture of iron(III) oxide and aluminium powder is placed in a test-tube standing in a tray of sand. It is ignited using a fuse of magnesium ribbon and a spectacular exothermic reaction follows producing molten iron.

Apparatus

▼   Several heat-proof mats to protect the bench.

▼   Small bucket of sand (a metal tray, biscuit tin or catering size coffee tin would do instead).

▼   One 16 mm x 150 mm test-tube. This will be destroyed so others will be needed if the demonstration is to be repeated.

▼   Safety screens.

▼   Access to oven set at between 75 °C and 100 °C.

▼   Desiccator.

▼   A magnet.

Chemicals

The quantities given are for one demonstration.

▼   11 g of fine aluminium powder.

▼   32 g of powdered iron(III) oxide. Precipitated 'red iron oxide, 85 % Fe_2O_3'.

▼   About 6 cm of magnesium ribbon.

▼   A little **magnesium powder** (optional).

Method

Before the demonstration

Twenty four hours before the demonstration, weigh out the aluminium and the iron oxide separately and place them in the oven. Shortly before the demonstration mix the two powders thoroughly and return the mixture to the oven or place it in a desiccator.

   Protect the demonstration bench with heat-proof mats and place the container of sand in the middle of these. Alternatively, consider doing the demonstration outside.

The demonstration

Fill the test-tube with the stoichiometric mixture of aluminium and iron oxide. Tap

THE ROYAL
SOCIETY OF
CHEMISTRY

the tube to ensure good packing. Leave about 2 mm space at the top. Add about 1 mm depth of magnesium powder and mix this into the top millimetre of the mixture. Fill the remaining space with magnesium powder but do not mix this with the powders. Take about a 6 cm length of magnesium ribbon and straighten it out. Insert

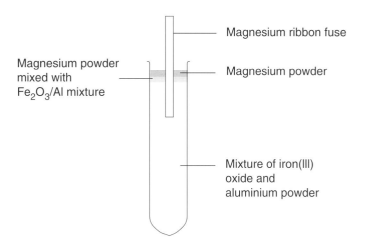

Fig. 1 The reaction container

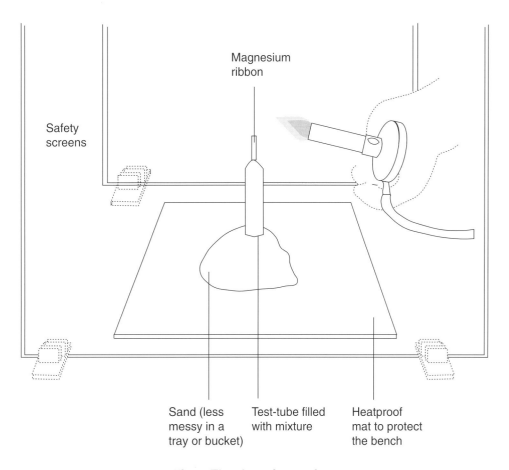

Fig. 2 The thermit reaction

1–2 cm of the ribbon into the mixture in the test-tube leaving the remaining 4 cm sticking out to act as a fuse *(Fig. 1)*. Place the tube in the container of sand so that about two-thirds of the tube protrudes. Set up safety screens to protect the audience and the demonstrator and ensure that the audience is at least 2 m from the test-tube *(Fig. 2)*.

NB: A fuse longer than that described tends to 'droop' when ignited and the burning end may break off and fail to ignite the mixture. The length suggested leaves ample time for retreating to a safe distance!

Using a roaring Bunsen flame, light the tip of the magnesium fuse. Stand well back as soon as it is lit. The magnesium will burn down to the mixture and ignite it. It will glow spectacularly and shoot out sparks leaving red hot molten iron. The test-tube will melt. Allow the experiment to cool (this will take several minutes) and break away the remains of the tube. Examine the lump of iron that remains and show that it is attracted to the magnet. Show that neither the iron oxide nor the aluminium powder is magnetic.

Note. If the mixture fails to ignite, take great care when approaching it and do not touch it unless you are absolutely certain that it has gone out completely. The mixture has been known to ignite some minutes after apparently failing. If in doubt, pour sand over the whole test-tube and leave it for several minutes.

Teaching tips

The energy level diagram for the reaction could be drawn and the idea of activation energy (provided by the burning magnesium) could be introduced.

This reaction was used to provide molten iron to weld railway lines. This is shown in a sequence from the Chemistry in Action video on Aluminium (Granada) which could be shown after the demonstration. The video also shows a thermit reaction in the laboratory.

Theory

The reaction is:

$$Fe_2O_3(s) + 2Al(s) \rightarrow Al_2O_3(s) + 2Fe(s) \quad \Delta H = -825 \text{ kJ mol}^{-1}$$

Extensions

If the reaction is carried out in a fireclay crucible rather than a test-tube, the molten iron can be poured into another crucible.

Further details

A variety of 'starter mixtures' has been suggested for this reaction. These are placed around the base of the magnesium fuse or used instead of a fuse and help to ensure that the reaction will ignite. The method suggested here seems to be reliable and will work well even without the extra magnesium powder provided that the reaction mixture is dry. CLEAPSS Hazcards suggests (on the barium peroxide card) a mixture of barium peroxide and magnesium powder. Several other mixtures are listed in *Tested demonstrations in chemistry* by N. H. Alyea and F. B. Dutton, *J. Chem. Ed.*, (6th edition) , 1965. None of these appears to be necessary, however and they all tend to obscure the simplicity of the reaction.

Safety

Wear eye protection.

It would be a sensible precaution to have a sand bucket and a fire extinguisher on hand but not a water based type.

It is the responsibility of teachers doing this demonstration to carry out an appropriate risk assessment.

THE ROYAL
SOCIETY OF
CHEMISTRY

75. The reaction of magnesium with steam

Topic

Reactions of metals, reactivity series.

Timing

About 5 min.

Level

Pre-16.

Description

Burning magnesium ribbon is plunged into the steam above boiling water in a conical flask. In the first method, the hydrogen that is formed is allowed to burn at the mouth of the flask. In the second method, the hydrogen is collected over water and tested with a lighted spill.

Apparatus

Method 1
▼   Bunsen burner, tripod and gauze.

▼   Tongs.

▼   One 250 cm³ conical flask.

Method 2
▼   Bunsen burner, tripod and gauze.

▼   One 1 dm³ conical flask with a one-holed rubber bung to fit.

▼   Glass trough or washing up bowl.

▼   One boiling tube.

▼   One short length of glass tube of approximately 1 cm diameter.

▼   About half a metre of rubber tubing.

▼   Wooden spills.

Chemicals

For both methods, the quantities given are for one demonstration.

▼   About 45 cm of magnesium ribbon.

▼   A little universal indicator liquid with appropriate colour chart.

Method

Before the demonstration
For method 2
Enlarge the hole in the rubber bung so that it will take a piece of glass tubing of diameter about 1 cm. Attach about half a metre of rubber delivery tube to this glass tube. This will be of similar bore to the tubing used for a Bunsen burner. The reason for this unusually wide tubing is so that it can cope with the rapid evolution of hydrogen that occurs in this demonstration.

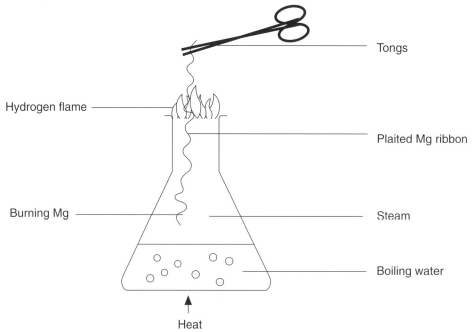

Fig. 1 Combustion of hydrogen

The demonstration
Method 1

Stand the 250 cm³ conical flask on the tripod and clamp its neck to steady it. Place about 50 cm³ of water in the flask. Bring this to the boil and allow it to boil for at least five minutes to displace all the air from the flask and replace it with steam. Take three 15 cm lengths of magnesium ribbon and twist them together to form a length of

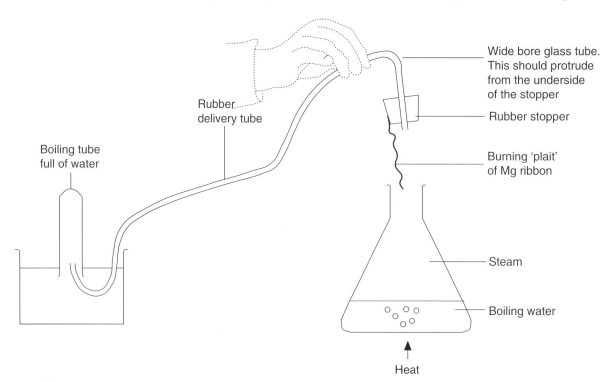

Fig. 2 Collection of hydrogen

THE ROYAL
SOCIETY OF
CHEMISTRY

plaited ribbon of the same length. This is more rigid than a single strand and can therefore be manoeuvred more easily when held in a pair of tongs. Take care that the ribbon does not break during plaiting. Leave the Bunsen burner on, boiling the water. Holding the plaited magnesium ribbon in tongs by one end, light the other end in the Bunsen flame (a second Bunsen burner may be helpful) and hold the burning end in the steam inside the flask (*Fig. 1*). Avoid looking directly at the burning ribbon. The ribbon will continue to glow brightly, forming hydrogen by reaction with steam. This ignites and burns at the mouth of the flask with a slightly yellowish flame. The magnesium oxide falls into the water and a little dissolves. Turn off the Bunsen burner and add a few drops of universal indicator to the water. It will be significantly alkaline due to dissolved magnesium hydroxide.

Method 2

Stand the 1 dm³ conical flask on the tripod and clamp its neck to steady it. Place about 200 cm³ of water in the flask. Bring this to the boil and allow it to boil for at least five minutes to displace all the air from the flask and replace it with steam. Plait the magnesium as described above and attach it to the underside of the bung on the wide bore delivery tube. The easiest way to do this is to cut a small slit in the rubber with a scalpel and insert one end of the plaited ribbon into the slit.

   Fill a trough with water and clamp a boiling tube full of water in an inverted position with its mouth under water. Place the free end of the rubber delivery tube in the mouth of the boiling tube. Clamp the delivery tube if necessary to prevent it coming out of the mouth of the boiling tube as the other end, attached to the bung, is moved (*Fig. 2*).

   Leave the Bunsen burner on, boiling the water. Light the end of the plaited magnesium ribbon and lower it into the steam in the flask until the bung is fitted into the mouth of the flask. The magnesium will continue to glow brightly in the steam, forming **hydrogen**. This will be forced along the delivery tube and some will be collected in the boiling tube, although much will overflow. Remove the bung and delivery tube from the flask to prevent suck-back and test the gas in the boiling tube with a lighted spill. It will 'pop' showing it to be hydrogen. The magnesium oxide will have fallen into the water and a little will have dissolved. Turn off the Bunsen burner and add a few drops of universal indicator to the water. It will be significantly alkaline due to dissolved magnesium hydroxide.

Visual tips

The hydrogen flame in method 1 would be more easily seen in a slightly darkened room.

Teaching tips

Do not allow the burning magnesium to touch the side of the flask. This can be a difficult task if you are dazzled by its flame. Wearing sunglasses might help. Compare the reaction of magnesium with steam with its reaction with cold water using the apparatus shown in *Fig. 3*. Very small bubbles will be seen on the surface of the magnesium but it will take several days before a significant volume can be collected.

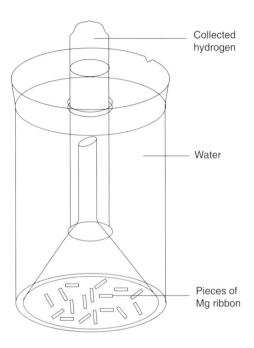

Fig. 3 Reaction of magnesium with cold water

Theory

The reaction is

$$Mg(s) + H_2O(g) \rightarrow MgO(s) + H_2(g)$$

Followed by

$$MgO(s) + H_2O(l) \rightarrow Mg(OH)_2(aq)$$

Extensions

Another method for this reaction *(Fig. 4)* is described in *Nuffield combined science teachers' guide II*, Sections 6 – 10. London: Longman/Penguin, 1970, 62. This may be suitable as a class experiment. The steam is generated by heating mineral wool soaked with water. However, the test-tubes (which must be of borosilicate glass) often crack and are ruined by the reaction of the hot magnesium with the glass.

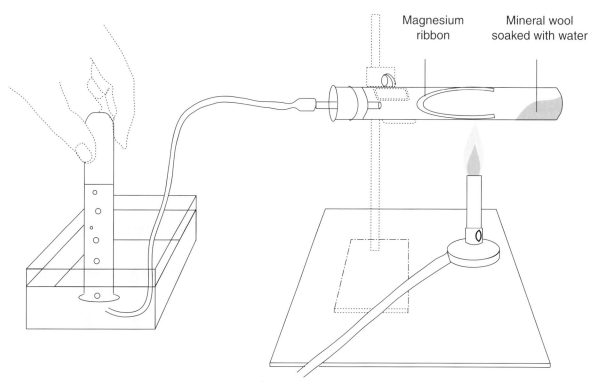

Fig. 4 An alternative method

Safety

Wear eye protection.

It is the responsibility of teachers doing this demonstration to carry out an appropriate risk assessment.

76. The reactions of chlorine, bromine and iodine with iron

Topic

Reactions of metals, reactions of halogens, periodicity.

Timing

About 10 min.

Level

Pre-16.

Description

Iron wool is heated in the presence of chlorine gas and the vapours of bromine and iodine. Exothermic reactions occur, the iron wool glows and iron(III) halides are formed. The vigour of the reactions corresponds to the order
chlorine > bromine > iodine, as would be predicted from the Periodic Table.

Apparatus

▼   Two boiling tubes.

▼   One reduction tube (ideally of the same size as the boiling tubes) with a one-holed rubber stopper fitted with a short length of glass tube.

If the chlorine is to be generated chemically:

▼   One 250 cm³ conical flask with a two-holed stopper to fit.

▼   A tap funnel.

▼   One piece of glass tubing with a right angled bend.

▼   Length of rubber delivery tubing.

▼   Access to fume cupboard.

Chemicals

The quantities given are for one demonstration.

▼   Three tufts of iron wool of mass about 1 g each. The finest grade is best since it gives the maximum surface area.

▼   About 0.5 cm³ of liquid **bromine**.

▼   About 0.5 g of **iodine**.

▼   About 100 cm³ of **1,1,1-trichloroethane** or other solvent for grease such as **hexane**.

▼   A little **silver nitrate** solution.

▼   A little deionised water.

▼   Access to a source of **chlorine** – either a cylinder or a chemical generator (use a tap funnel to drip **concentrated hydrochloric acid** onto solid potassium manganate(VII). If chlorine generated chemically is to be used, about 10 g of **potassium permanganate** (potassium manganate(VII), $KMnO_4$) and 100 cm³ of concentrated hydrochloric acid will be needed).

THE ROYAL
SOCIETY OF
CHEMISTRY

Method

Before the demonstration

Set up the chlorine generator in the fume cupboard if necessary.

Degrease the iron wool by dipping it in 1,1,1-trichloroethane or other grease solvent and then allow the solvent to evaporate.

The demonstration

Iodine

Place about 0.5 g of iodine in a boiling tube. Place a 1 g tuft of iron wool in the tube so that it is well spread out and almost fills the tube. Clamp the tube at about 45° to the horizontal in the fume cupboard. Heat the iodine gently with a Bunsen flame until it vaporises and purple iodine vapour rises up the tube to meet the iron wool. Then heat the iron wool. The wool will glow dully and the iron will become coated with red-brown iron(III) iodide, some of which may escape as 'smoke' from the top of the tube. When the reaction is complete, remove the iron wool from the tube with tweezers and dip it in a small beaker of deionised water to dissolve some of the product. Test this solution with silver nitrate solution; a yellow precipitate confirms that iodide ions are present.

NB: If the iron wool is removed too soon, the reaction may still be going on and the iron may begin to glow again. This may well be a reaction with oxygen in the air rather than with any remaining iodine vapour.

Bromine

Place about 0.5 cm³ of liquid bromine in a boiling tube. Place a 1 g tuft of iron wool in the tube so that it is well spread out and almost fills the tube. Clamp the tube at about 45° to the horizontal in the fume cupboard (Fig. 1). Heat the bromine gently with a Bunsen flame until it vaporises and brown bromine vapour rises up the tube to meet the iron wool. Then heat the iron wool. The wool will glow and the iron will become coated with yellow-brown iron(III) bromide, some of which may escape as 'smoke' from the top of the tube. When the reaction is complete, remove the remaining iron wool from the tube with tweezers and dip it in a small beaker of deionized water to dissolve some of the product. Test this solution with silver nitrate solution; a cream precipitate confirms that bromide ions are present.

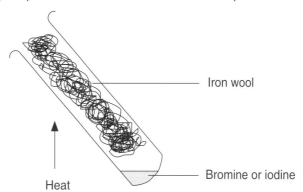

Iron wool

Bromine or iodine

Heat

Fig.1 Reaction of iron with bromine or iodine

Chlorine

Connect the reduction tube to the chlorine source with rubber tubing. Place a 1 g tuft of iron wool in the tube so that it is well spread out and almost fills the tube. Clamp the tube horizontally in the fume cupboard. Pass a slow stream of chlorine over the iron wool. The wool may ignite without heating. If it does not, heat the wool gently nearest to the generator with a Bunsen flame until it does ignite. A vigorous reaction will occur and the glow will spread through the iron wool producing clouds of brown

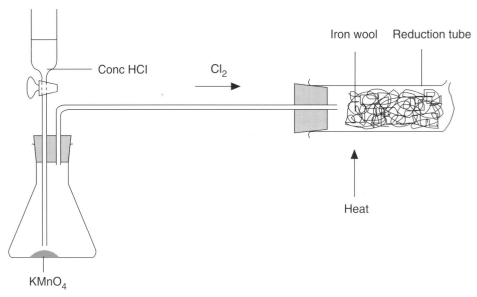

Chlorine generator

Fig. 2 Reaction of iron with chlorine

iron(III) chloride, some of which will emerge from the hole in the reduction tube. Little or no iron will remain. After the reaction has finished, turn off the chlorine and, when the tube is cool, rinse it with deionised water to dissolve some of the product. Test the resulting solution with silver nitrate solution; a white precipitate shows that chloride ions are present.

Teaching tips

Do the reactions in the order given above so that the most reactive is last. Point out the increase in reactivity of the halogens on ascending the group.

Theory

The reactions are (using X to represent any halogen):

$$2Fe(s) + 3X_2(s, l, or g) \rightarrow 2FeX_3(s)$$

Extensions

The residues can be tested to confirm the presence of iron(III) ions if desired and if the students are familiar with the tests. Dissolve a little of the residues in deionised water and add a few drops of either potassium (or ammonium) thiocyanate solution or potassium hexacyanoferrate(II) (potassium ferrocyanide) solution. These will give a blood red colour and a deep blue colour (Prussian blue) respectively.

Further details

Take care if transferring bromine using a dropping pipette. It has a high density and a high vapour pressure which tend to eject the bromine from the pipette.

It is difficult to predict whether or not the iron will ignite in chlorine without heating. It appears to depend on its surface area, cleanliness and the rate of flow of the chlorine. Prior heating in a stream of hydrogen does not appear to make any difference.

Safety

Wear eye protection.

It is the responsibility of teachers doing this demonstration to carry out an appropriate risk assessment.

THE ROYAL
SOCIETY OF
CHEMISTRY

77. The reactions of aluminium with chlorine, bromine and iodine

Topic

Reactions of metals, reactions of halogens or general interest.

Timing

Less than 5 min each.

Level

Pre-16, any for general interest.

Description

A little water is added to a mixture of powdered iodine and aluminium powder. A spectacular exothermic reaction occurs producing flames and clouds of iodine vapour as well as aluminium iodide.

Small pieces of torn aluminium foil are floated on the surface of liquid bromine. After about a minute flames and flashes of light can be seen along with a white 'smoke' of aluminium bromide.

Chlorine is passed over a crumpled piece of aluminium foil in a combustion tube. Gentle heating starts an exothermic reaction which causes the aluminium to glow white, spreading along the foil producing white aluminium chloride, some of which condenses on the combustion tube and some of which escapes as white 'smoke'.

Apparatus

▼ Mortar and pestle.

▼ Heat-proof mat.

▼ Large watch glass (approximately 10 cm in diameter).

▼ Reduction tube with one-holed rubber stopper fitted with a short length of glass tube.

If the chlorine is to be generated chemically:

▼ One 250 cm³ conical flask with a two-holed stopper to fit.

▼ A tap funnel.

▼ One piece of glass tubing with a right angled bend.

▼ Length of rubber delivery tubing.

▼ Access to fume cupboard.

THE ROYAL
SOCIETY OF
CHEMISTRY

Chemicals

The quantities given are for one demonstration.

▼   0.3 g of aluminium powder.

▼   2 g of **iodine**.

▼   1 cm³ of **bromine**.

▼   A few cm² of aluminium foil.

▼   A little **silver nitrate** solution.

▼   A little deionised water.

▼   Access to a source of **chlorine** – either a cylinder or a chemical generator (use a tap funnel to drip **concentrated hydrochloric acid** onto solid **potassium permanganate** (potassium manganate(VII)). If chemically generated chlorine is to be used, about 10 g of potassium manganate(VII) (potassium permanganate, $KMnO_4$) and 100 cm³ of concentrated hydrochloric acid will be needed.

Method

Before the demonstration
Set up the chlorine generator (if required).

The demonstration
Aluminium and chlorine
Crumple a piece of aluminium foil about 10 cm x 5 cm and place it in a reduction tube. Clamp the reduction tube horizontally in a fume cupboard. Connect the combustion tube to the chlorine cylinder or generator in a fume cupboard. Pass a gentle stream of chlorine over the aluminium and heat the aluminium gently with a Bunsen flame at the end nearest to the chlorine generator. The aluminium will begin to glow white. Remove the Bunsen burner and the glow will spread along the aluminium. A white 'smoke' of aluminium chloride will be formed, some of it condensing on the walls of the reduction tube and some streaming through the hole in the end. Test a little of this white powder with moist universal indicator paper. It will be acidic. Scrape a little of the powder off with a spatula, dissolve it in deionised water and test with silver nitrate solution. A white precipitate will form, showing the presence of chloride ions.

Aluminium and bromine
Tear a few pieces of aluminium foil about 2 mm x 2 mm. Break open a 1 cm³ ampoule of bromine and pour the bromine onto a watchglass in the fume cupboard. Sprinkle the pieces of foil onto the surface of the bromine. Within a minute, flashes and flames are seen and a white 'smoke' of aluminium bromide is formed. If desired, hold a watchglass over the reaction to condense some of the 'smoke' to be tested with silver nitrate.

Aluminium and iodine
Weigh out 2 g of iodine, which should be dry, and grind it to a powder if necessary using a mortar and pestle. Add to this 0.3 g of aluminium powder and mix well with a spatula, but do not grind them together. Pour the mixture onto a heatproof mat or a metal tray, such as a tin lid, in a fume cupboard. Use a dropping pipette to place a few drops of water on the mixture. After a lag of about 10 seconds, the water begins to steam and clouds of purple iodine vapour are given off indicating that an exothermic reaction has started. After a few more seconds, sparks are given off and

the mixture bursts into flame. White aluminium iodide remains. Dissolve a little of this in deionised water, filter to remove unreacted aluminium and test with silver nitrate solution. A yellow precipitate indicates that iodide ions are present.

Visual tips

The bromine reaction would look spectacular in a dark room.

Teaching tips

The apparent order of vigour of these reactions is not in the order that would be predicted from the Periodic Table – *ie* chlorine > bromine > iodine. Point out that this is due to the different physical states of the halogens, the different surface areas of the metal and the poor contact between solid iodine and solid aluminium. It is also worth noting the oxide film that occurs on aluminium.

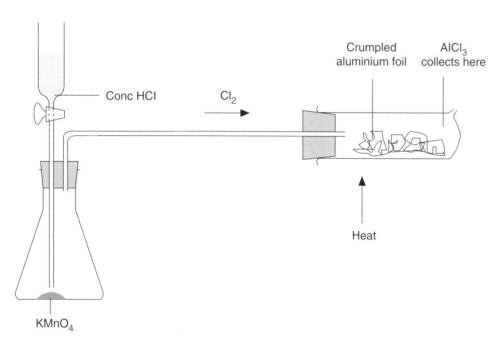

Reaction of aluminium with chlorine

Theory

The reactions are (using X to represent any halogen):

$$2Al(s) + 3X_2(s, l \text{ or } g) \rightarrow 2AlX_3(s)$$

In the iodine reaction, the function of the water is to dissolve a little iodine and allow the reactants to come together.

Extensions

The reaction of zinc and iodine is similar to the reaction of aluminium and iodine. Powder about 7.5 g of iodine as above and mix thoroughly with about 2 g of powdered zinc in a boiling tube. These are approximately the stoichiometric quantities. Do not grind the zinc and iodine together in a mortar and pestle because the reaction may start during the grinding. In a fume cupboard, add 2 or 3 cm³ of water from a teat pipette. Immediately, an exothermic reaction occurs producing clouds of iodine vapour and making the boiling tube very hot.

THE ROYAL
SOCIETY OF
CHEMISTRY

Further details

A simpler method of demonstrating the reaction of aluminium with chlorine is to use tongs to lower a heated piece of aluminium foil into a gas jar of chlorine. The method described above has the advantage of not confusing the reaction with air with that with chlorine.

The method used for the reaction between aluminium and chlorine can be modified to collect the product more efficiently. Slight modifications to the apparatus allow the reactions with chlorine or other elements to be demonstrated and the products collected. The formulae of the chlorides can be determined by titration with silver nitrate solution using potassium chromate as an indicator. These experiments are described in *Nuffield advanced science: chemistry students' workbook I* (1st edn), p22. London: Longman, 1970.

Some teachers may wish to demonstrate the aluminium and iodine reaction outdoors because a lot of iodine vapour is produced which could stain the inside of the fume cupboard.

Safety

Wear eye protection.

It is the responsibility of teachers doing this demonstration to carry out an appropriate risk assessment.

THE ROYAL
SOCIETY OF
CHEMISTRY

78. Following the reaction of sodium thiosulphate and acid using a colorimeter

Topic

Reaction rates.

Timing

About 10 min.

Level

Pre-16.

Description

The reaction of sodium thiosulphate with acid produces a precipitate of colloidal sulphur. The reaction can be followed on a simple colorimeter by monitoring the amount of light transmitted by the reaction mixture.

Apparatus

▼ A simple colorimeter such as a Griffin and George environmental comparator. Most types would be suitable.

▼ Sample tube for the colorimeter.

▼ Small measuring cylinder and/or graduated pipette – the exact size depends on the size of the sample tubes.

▼ One boiling tube.

Chemicals

The quantities given are for one demonstration, but both quantities and concentrations may need adjusting depending on the colorimeter used.

▼ 100 cm^3 of 0.05 mol dm^{-3} sodium thiosulphate.

▼ 10 cm^3 of 1 mol dm^{-3} **hydrochloric acid**.

Method

Before the demonstration
Set up and calibrate the colorimeter according to the manufacturer's instructions *ie* set the zero and 100 % readings.

The demonstration
Almost fill the boiling tube with sodium thiosulphate solution. Add about 1 cm^3 of hydrochloric acid, stopper and invert the tube to mix the reactants. After about 30 seconds, a cloudy white precipitate of sulphur will become apparent. This will get thicker until the solution becomes opaque. Reaction appears to be complete after about 2 minutes. **Sulphur dioxide** can be smelt.

Now repeat the reaction in the sample tube of the colorimeter, placing the tube in the colorimeter immediately after mixing the reactants. Take readings of light transmitted at suitable intervals (about every ten seconds) for two minutes.

The exact volumes of solutions to be mixed will depend on the size of the sample cell. They are not critical, but try to keep the proportions as described above, *ie* about 50 of thiosulphate to 1 of acid.

Teaching tips

This would be a good demonstration to follow a class experiment on the same reaction by the 'disappearing cross' method as described in, for example, *Revised nuffield chemistry, teachers' guide II*, p229. London: Longman, 1970. It provides a simple introduction to the idea of following the course of a reaction by colorimetry. Many teachers may prefer to stress the technique of colorimetry rather than the actual results since these are not typical of those students will meet at this level.

One student could be asked to take the readings and another to note them down on the blackboard or the students could have ready prepared axes and plot the graph as they watch.

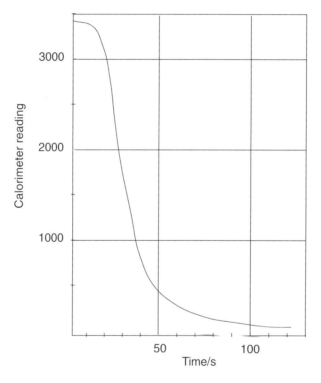

Typical set of results

This would be an ideal experiment for computer interfacing. The graph could then be plotted 'on line' and hard copies run off for the students. Many colorimeters have terminals suitable for connection to a microcomputer. R. Edwards, *Interfacing chemistry experiments*. London: RSC, 1993 gives helpful information on computer interfacing.

Theory

The reaction is:

$$Na_2S_2O_3(aq) + 2HCl(aq) \rightarrow 2NaCl(aq) + S(s) + SO_2(aq) + H_2O(l)$$

Extensions

Try varying the concentrations of the acid and the thiosulphate.

THE ROYAL
SOCIETY OF
CHEMISTRY

Further details

The kinetics of this reaction are complex. There is a brief discussion of them in M. A. Atherton and J. L Lawrence, *An experimental introduction to reaction kinetics*, p157. London: Longman, 1970. A rate law of the form

$$\text{Rate of reaction} = \frac{k' [H^+]}{1 + k'' [H^+]} \times [S_2O_3^{2-}]$$

is suggested.

The rate curve obtained in this experiment (*see Figure*) is not of the shape expected for a simple rate expression. This is presumably due to the complex kinetics and the fact that the colorimeter reading is not proportional to concentration.

The construction of a home-made colorimeter, suitable for this experiment, based on an Energy Studies Measurement Instrument (ESMI) available from British Gas is described in *Sch. Sci. Rev.*, **69** (246), 1987, 91.

Safety

Wear eye protection.

Dispose of the opaque residue as soon as possible by flushing down the sink with plenty of water. Sulphur dioxide is toxic by inhalation and can affect asthma sufferers adversely.

It is the responsibility of teachers doing this demonstration to carry out an appropriate risk assessment.

THE ROYAL
SOCIETY OF
CHEMISTRY

79. The fountain experiment

Topic

Gases – solubility and acidity/alkalinity.

Timing

About 5 min per gas.

Level

Introductory chemistry.

Description

A 1 or 2 dm³ flask is filled with a soluble gas such as ammonia, hydrogen chloride or sulphur dioxide. A length of glass tubing, the upper end of which is drawn out into a jet, is fitted through the stopper and the lower end placed into a beaker of water containing indicator. As the gas dissolves in the water, a partial vacuum is formed in the flask which sucks more water inside as a fountain. In the fountain, the indicator changes colour.

Apparatus

▼   One 1 dm³ or 2 dm³ flask. Any shape will work, although a round one looks most attractive.

▼   A stopper to fit the flask.

▼   A two-holed rubber stopper to fit the flask.

▼   A length of glass tube, one end of which is drawn out into a jet – a 1 cm³ pipette would do. The length should be such that when the tube is fitted through the stopper, the jet is near the centre of the flask and about 20 cm of tube protrudes through the stopper.

▼   One 10 cm³ or 20 cm³ plastic syringe.

▼   Trough or large beaker which can hold more water than the flask in which the fountain is to be produced.

▼   Access to a fume cupboard.

For preparing the gases that are to be demonstrated :

Ammonia

▼   One boiling tube fitted with a one-holed rubber stopper holding a drying tube.

▼   One-holed rubber stopper to fit the drying tube, fitted with a short length of glass delivery tube.

▼   Length of rubber delivery tube.

▼   Bunsen burner or hair drier.

Hydrogen chloride

▼   One 250 cm³ conical flask with a two-holed rubber stopper to fit.

▼   Tap funnel.

THE ROYAL
SOCIETY OF
CHEMISTRY

▼   Short length of glass tube with a right angled bend.

▼   Length of rubber delivery tube.

Sulphur dioxide

▼   One 250 cm³ conical flask with a two-holed rubber stopper to fit.

▼   Tap funnel.

▼   Short length of glass tube with a right angled bend.

▼   Length of rubber delivery tube.

▼   Bunsen burner, tripod and gauze.

Chemicals

The quantities given are for one demonstration.

▼   A few cm³ of universal indicator solution (or other indicator if preferred).

Chemicals required to prepare the gases that are to be demonstrated. Sulphur dioxide may be obtained from a cylinder.

Ammonia

▼   About 10 cm³ of **880 ammonia solution**.

▼   About 10 g of **potassium hydroxide** pellets (enough to half fill the drying tube).

Hydrogen chloride

▼   About 100 cm³ of **concentrated sulphuric acid**.

▼   About 40 g of sodium chloride (common salt, NaCl).

Sulphur dioxide

▼   About 100 cm³ of **concentrated sulphuric acid**.

▼   About 50 g of copper turnings.

Method

Before the demonstration

Ensure that the large flask is completely dry; if necessary rinse with propanone and allow to evaporate.

Take the two-holed stopper that fits the large flask and fit the glass tube, which has been drawn out into a jet, through one of the holes. The jet should be positioned so that it is near the centre of the flask when the stopper is in place and so that about 20 cm of tube protrudes from the stopper. Fit the nozzle of the plastic syringe into the other hole. Fill the syringe with water (*Fig. 4*).

Fill the trough or large beaker with enough water to fill the fountain flask and add enough indicator to give an easily visible colour – this will be much more than is used for a titration. If necessary, add a little dilute acid or alkali so that the indicator starts in its acid colour for ammonia and its alkaline colour for the other two gases.

Working in a fume cupboard, prepare the gas (as described below) and fill and stopper the flask. Some teachers will prefer to do this in the presence of the class.

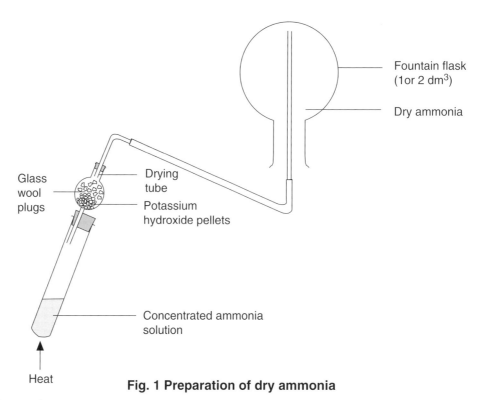

Fig. 1 Preparation of dry ammonia

Ammonia

Set up the boiling tube and drying tube as shown in *Fig. 1*. Half fill the drying tube with pellets of potassium hydroxide and half fill the boiling tube with 880 ammonia solution. Clamp the large flask in an inverted position and arrange a delivery tube for downward displacement of air because ammonia is less dense than air. Warm the ammonia solution gently with a small Bunsen flame (or a hair drier) for a few minutes. Ammonia gas comes out of solution and is dried by the potassium hydroxide. Confirm that the flask is full by holding moist indicator paper around the base of the flask and looking for an alkaline reaction. If in doubt continue filling for a little longer.

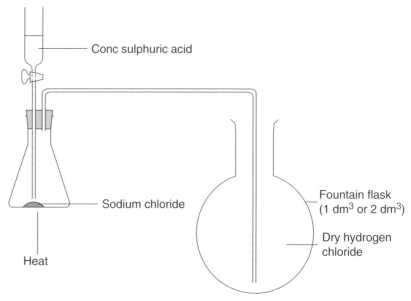

Fig. 2 Preparation of dry hydrogen chloride

Hydrogen chloride

Set up the conical flask, tap funnel and delivery tube as shown in *Fig. 2*. Clamp the large flask upright and arrange a delivery tube for upward displacement of air, because hydrogen chloride is denser than air. Place about 40 g of sodium chloride in the conical flask and half fill the tap funnel with concentrated sulphuric acid. Drip the acid on to the sodium chloride to generate hydrogen chloride gas and fill the large flask for a few minutes. Steamy fumes will be seen around the mouth of the flask when it is full and this can be confirmed by testing around the mouth of the flask with moist indicator paper and looking for an acid reaction. If in doubt continue filling for a little longer.

Sulphur dioxide

Set up the conical flask, tap funnel and delivery tube on a tripod and gauze above a Bunsen burner as shown in *Fig. 3*. Clamp the large flask upright and arrange a delivery tube for upward displacement of air, because sulphur dioxide is denser than

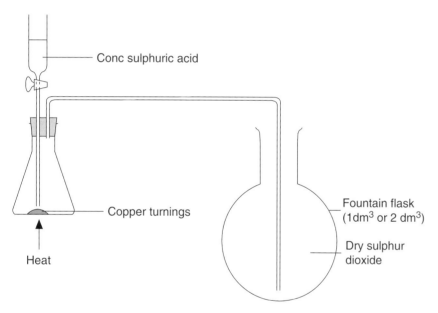

Fig. 3 Preparation of dry sulphur dioxide

air. Place about 50 g of copper turnings in the conical flask and half fill the tap funnel with concentrated sulphuric acid. Drip about 50 cm³ of acid onto the copper and heat gently until evolution of sulphur dioxide starts. Continue to heat as necessary to maintain a steady evolution of gas. Fill the large flask for a few minutes. Test around the mouth of the flask with moist indicator paper, looking for an acid reaction to confirm when the flask is full. If in doubt continue filling for a little longer.

The demonstration

Remove the stopper from the gas-filled flask and quickly fit the stopper with jet and syringe. Clamp (or get an assistant to hold) the flask over the trough or beaker of water so that the protruding glass tube is well below the water level. If clamping, bear in mind that the flask will be heavy when filled with water so take care that it will not overbalance. Use the syringe to squirt a few cm³ of water into the flask and swirl gently to dissolve some of the gas. As the gas dissolves, a partial vacuum will form inside the flask and the external air pressure will force water up the tube and through the jet forming a fountain (Fig. 4). As the water emerges from the jet, it is exposed to the acidic or alkaline gas and the indicator will change colour. The fountain will

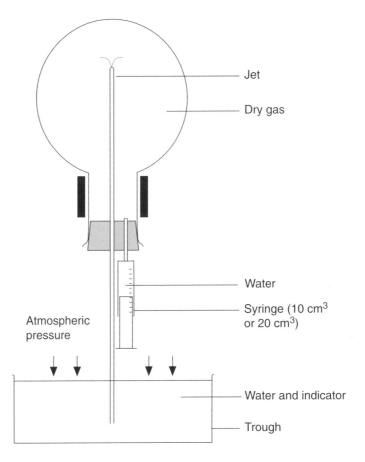

Fig. 4 The fountain apparatus

continue for some minutes depending on the size of the flask and the width of the jet. When the fountain finishes, a bubble of gas will remain. This is air and its volume gives an indication of how well the flask was filled orignally.

Visual tips

A white background is ideal.

Teaching tips

The dissolution of sulphur dioxide in water, forming sulphuric(IV) acid (sulphurous acid) is a step in the formation of acid rain.

Theory

The reactions by which the gases are prepared are:

$NaCl(s) + H_2SO_4(l) \rightarrow NaHSO_4(s) + HCl(g)$

$Cu(s) + 2H_2SO_4(l) \rightarrow CuSO_4(aq) + 2H_2O(l) + SO_2(g)$

A side reaction produces copper(I) sulphide which is responsible for the black colour of the reaction mixture.

When sulphur dioxide reacts with water, the following reaction takes place, producing sulphuric(IV) acid:

$SO_2(g) + H_2O(l) \rightarrow H_2SO_3(aq)$

THE ROYAL
SOCIETY OF
CHEMISTRY

Extensions

It is possible to set up a chemiluminescent fountain. Attach to the base of the jet tube a Y-piece, one end of which dips into a beaker containing an aqueous solution of luminol and sodium hydroxide while the other dips into a solution of bleach. See demonstration 6 for details of the solutions. The two solutions are sucked into the flask, mix in the jet and react giving out blue light which looks spectacular in a darkened room.

An alternative way to start the fountain is to pour liquid nitrogen (if available) over the fountain flask. This cools the gas which contracts and reduces the pressure inside the flask, sucking in the water. Pouring a mixture of dry ice and ethanol over the flask may also start the fountain as may ethoxyethane (diethyl ether, ether), which cools the flask as it evaporates, but these methods are less satisfactory from a safety point of view.

Another method is to warm the flask gently with a small Bunsen flame or hair drier until a few bubbles of gas emerge from the bottom of the jet tube. Allow the flask to cool; this will suck water up the tube, starting the fountain.

Other gases of similar solubility can be used.

Further details

The solubilities of the gases used are:

ammonia	1 300
hydrogen chloride	506
sulphur dioxide	80

These figures represent the volume of gas (at 273 K and 100 kPa) that will dissolve in one volume of water.

Safety

Wear eye protection.

It is the responsibility of teachers doing this demonstration to carry out an appropriate risk assessment.

THE ROYAL
SOCIETY OF
CHEMISTRY

80. The preparation and properties of nitrogen(I) oxide

Topic

Oxides of nitrogen, Gay-Lussac's Law.

Timing

About 20 min but the demonstration of solubility takes longer – about an hour.

Level

Pre-16 or post-16.

Description

Nitrogen(I) oxide (dinitrogen monoxide, dinitrogen oxide, nitrous oxide, N_2O) is prepared by the reaction of sodium nitrate and ammonium sulphate. It is shown to relight a glowing splint and to be somewhat soluble in water. The gas is passed over heated copper to leave nitrogen; there is no volume change.

Apparatus

▼   Boiling tube with one-holed rubber stopper fitted with a short length of glass tube.

▼   A length of rubber delivery tube.

▼   Glass trough or washing-up bowl.

▼   Two or three test-tubes (with stoppers) in which to collect the gas.

▼   Two 100 cm^3 gas syringes.

▼   A 15 cm length of Pyrex tube, about 8 mm in diameter.

▼   Bunsen burner.

▼   Heat-proof mat.

▼   A three-way stop cock (optional).

▼   Wooden tapers.

Chemicals

The quantities given are for one demonstration.

▼   5 g of sodium nitrate (sodium nitrate (V), $NaNO_3$)

▼   4 g of ammonium sulphate (($NH_4)_2)SO_4$).

▼   A few grams of copper turnings.

▼   Access to a cylinder of nitrogen.

Method

Before the demonstration

Set up the gas syringes so that they are clamped about 20 cm above the bench and are connected by a 15 cm long piece of Pyrex tube and a three-way stop cock (optional) (*Fig. 1*). Pack the Pyrex tube with copper turnings.

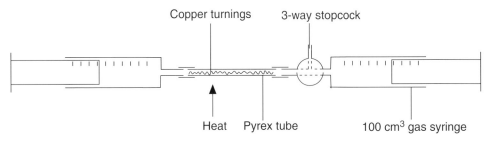

Copper turnings   3-way stopcock

Heat   Pyrex tube   100 cm^3 gas syringe

Fig. 1 Preparing nitrogen (I) oxide

The demonstration

Half fill the trough with water and fill with water as many test-tubes as will be
required. Weigh out 5 g of sodium nitrate and 4 g of ammonium sulphate (these are
approximately the stoichiometric quantities). Mix them thoroughly and place in a
boiling tube. Fit the stopper and delivery tube and clamp the boiling tube
horizontally with the end of the delivery tube in the trough of water. Heat the
mixture. This will 'melt' and gas will be given off. Collect several test-tubes full of the
gas and discard the first two as they will contain air. White fumes will be seen inside
the boiling tube and possibly in the collected gas. These are presumably unreacted
starting material. They will dissolve in water, and a test-tube containing white fumes
clears if a little water is left in it and shaken up.

If brown fumes are seen, remove the delivery tube, stop heating and allow the
apparatus to cool.

Show that the gas relights a glowing taper.

Clamp one test-tube of the gas vertically so that its open end is under water. Leave
this; the water level in the tube will rise showing that the gas is somewhat soluble.
The gas will slowly rise about 2 cm over about one hour. Use water that is as cold as
possible – nitrogen(I) oxide is about twice as soluble at 0 °C as at 20 °C. If desired,
clamp a tube of air next to the tube of nitrogen(I) oxide; no change of level will be
seen.

One teacher suggested using an inverted funnel, the stem of which is closed with
a short length of rubber tube and a screw clip instead of an inverted test-tube (*Fig. 2*).
This increases the surface area of gas exposed to the water and speeds up the rise of
the water level.

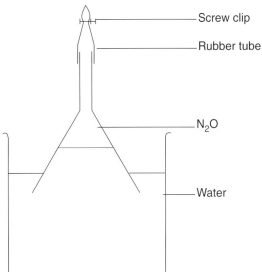

Screw clip

Rubber tube

N$_2$O

Water

Fig. 2 Inverted funnel arrangement for fast nitrogen(I) oxide dissolution

Flush the syringes and Pyrex tube with nitrogen from a cylinder by filling each syringe several times and expelling the gas. Fill one of the syringes to the 80 cm^3 mark with nitrogen(I) oxide. Heat the copper with a Bunsen flame and pass the gas from syringe to syringe several times. Some of the copper turns black as copper oxide is formed. Return the gas to the original syringe, allow it to cool back to room temperature and read its volume. There will be no change. Remove the syringe containing the residual gas and fill two test-tubes with it by displacement of water. Show that this gas will extinguish both a glowing taper and a burning taper, suggesting that it is nitrogen.

Theory

The reactions by which the nitrogen(I) oxide is prepared are:

$$(NH_4)_2SO_4(s) + 2NaNO_3(s) \rightarrow 2NH_4NO_3(s) + Na_2SO_4(s)$$

followed by:

$$NH_4NO_3(s) \rightarrow N_2O(g) + 2H_2O(l)$$

Direct heating of **ammonium nitrate** is not recommended because the temperature must be controlled carefully to avoid the alternative, explosive, decomposition:

$$2NH_4NO_3(s) \rightarrow 2NO(g) + N_2(g) + 4H_2O(l)$$

that occurs at over 200 °C.

The reaction of nitrogen(I) oxide with copper is:

$$N_2O(g) + Cu(s) \rightarrow N_2(g) + CuO(s)$$

therefore no volume change is observed.

The glowing taper relights because of the decomposition of nitrogen(I) oxide into nitrogen and oxygen which takes place at over 500 °C:

$$2N_2O(g) \rightarrow 2N_2(g) + O_2(g)$$

Extensions

Many substances will continue to burn in nitrogen(I) oxide if they have been previously ignited in air. These include iron wool, charcoal and phosphorus. Burning sulphur will not continue to burn in nitrogen(I) oxide as it burns at a temperature lower than the decomposition temperature of nitrogen(I) oxide – which is 500 °C.

Further details

Nitrogen(I) oxide is 'laughing gas' which is still occasionally used as an anaesthetic. It is not recommended that its anaesthetic properties be tried as this may well tend to encourage potential substance abusers. Also, the gas is likely to contain some nitrogen dioxide as an impurity.

Potassium nitrate can be used as an alternative to sodium nitrate.

The relevant solubilities in water are:

nitrogen(I) oxide	1.3
oxygen	0.05
nitrogen	0.024

THE ROYAL
SOCIETY OF
CHEMISTRY

These figures represent the volume of gas (at 273 K and 100 kPa that will dissolve in one volume of water).

Safety

Wear eye protection.

It is the responsibility of teachers doing this demonstration to carry out an appropriate risk assessment.

THE ROYAL
SOCIETY OF
CHEMISTRY

81. The equilibrium between nitrogen dioxide and dinitrogen tetroxide

Topic

Equilibrium, Le Chatelier's principle.

Timing

About an hour.

Level

Post-16.

Description

A gas syringe is filled with an equilibrium mixture of brown nitrogen dioxide and colourless dinitrogen tetroxide. The position of the equilibrium can be gauged by the colour of the mixture and by its total volume. The effect of pressure can be demonstrated by compressing the mixture and observing the change in colour. The effect of temperature can be demonstrated by heating the mixture in a water bath and comparing the volume increase with that of a similar volume of air.

Apparatus

▼   Two 100 cm³ clear glass gas syringes (the experiment can be done with one).

▼   Boiling tube with a side arm with a one-holed rubber stopper to fit.

▼   Boiling tube with a one-holed rubber stopper to fit.

▼   Two short lengths of glass tubing.

▼   Length of rubber delivery tube.

▼   Short length of plastic tubing which is a very tight fit over the nozzles of the syringes.

▼   Two Hoffmann screw clips.

▼   One 400 cm³ beaker.

▼   One 2 dm³ beaker.

▼   One 0 –100 °C thermometer.

▼   Three-way stop cock or glass T-piece (optional).

▼   Bunsen burner, tripod and gauze.

▼   Access to fume cupboard.

Chemicals

The quantities given are for one demonstration.

▼   About 20 g of **lead nitrate** (Pb(NO₃)₂).

▼   About 400 cm³ of crushed ice.

▼   About 100 g of common salt (sodium chloride).

▼   A few drops of light lubricating oil such as sewing machine oil.

THE ROYAL
SOCIETY OF
CHEMISTRY

Method

Before the demonstration

Dry the lead nitrate overnight in an oven set at around 100 °C. Store the lead nitrate in a desiccator unless it can be used straight from the oven. In a fume cupboard, set up the apparatus for collection of dinitrogen tetroxide shown in *Fig. 1*. Keep the length of the rubber connecting tube as short as possible because **nitrogen dioxide** attacks rubber.

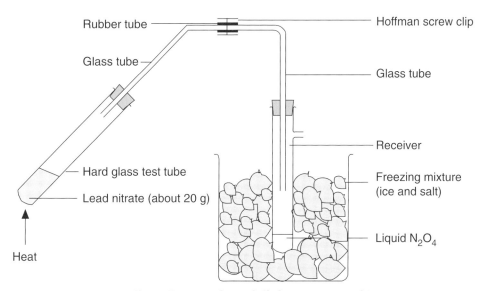

Fig. 1 Preparation of dinitrogen tetroxide

Lubricate the plunger of one of the syringes with a few drops of light oil such as sewing machine oil to prevent leaking under pressure. Fit a short length (about 5 cm) of transparent plastic tubing on to the syringe nozzle. It is vital that this is a tight fit, so select a piece that is just too narrow to fit when cold and immerse it in hot water to soften it before fitting. Test the tightness of the seal by closing the plastic tube with a screw clip when there is about 50 cm³ of air in the syringe. Push in the plunger to compress the gas as far as possible. The tubing should not come off the nozzle. This test is important to make sure that the tube will not come off the nozzle when the nitrogen dioxide gas is compressed.

Mix the salt with the crushed ice to produce a freezing mixture.

The demonstration

Working in a fume cupboard, heat the lead nitrate gently to decompose it to nitrogen dioxide. Heating too vigorously may decompose the gas into nitrogen monoxide. As the gas meets the cold wall of the side arm test-tube, it condenses as liquid dinitrogen tetroxide which may appear greenish due to dissolved water if the lead nitrate was not absolutely dry. When about 2 cm³ of liquid has been collected, stop heating, tighten the screw clip and remove the tube containing the lead nitrate. The liquid can be kept for some time in the freezing mixture (or in a freezer) so that this part of the experiment could be done before the demonstration if desired.

Connect the side arm of the tube containing the dinitrogen tetroxide to the plastic tube on the syringe nozzle via the three-way tap or T-piece. Partly fill the syringe with nitrogen dioxide gas by gently warming the boiling tube in the hand, a beaker of hot water or over a very small Bunsen flame. Flush the gas out into the fume cupboard and repeat the filling and flushing cycle two or three times to ensure that there is no air in the system. If a T-piece is used, a rubber-gloved finger can be used over the tube which is open to the air to control the filling and flushing process. Fill the syringe to the 50 cm³ mark with nitrogen dioxide, fit a screw clip over the plastic tube and tighten this to seal the gas in the syringe (*Fig. 2*).

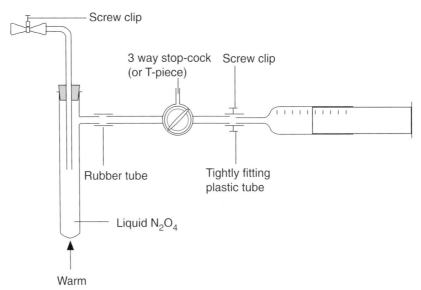

Fig. 2 Preparation of an equilibrium mixture of nitrogen dioxide and dinitrogen tetroxide

Effect of pressure on the equilibrium

In view of the audience, press in the plunger of the syringe as far as possible. The colour of the gas mixture will become darker as the concentration of the gas mixture increases. Within a few seconds it will become paler, as the position of the equilibrium responds to the increased pressure, and brown nitrogen dioxide is converted into colourless dinitrogen tetroxide. Release the plunger and the changes will reverse themselves. Pull the plunger out to reduce the pressure and note the colour changes.

Effect of temperature on the equilibrium

Fill a second syringe with 50 cm³ of air and seal with a screw clip in the same way as the first one. Clamp both syringes vertically in a 2 dm³ beaker of water so that they are immersed up to the 100 cm³ mark. Note the temperature and the readings of both syringes (which should be the same). Heat the water gently with a Bunsen burner and record the temperature and the volume of gases every 10 °C or so. Before taking each reading, remove the Bunsen burner and stir the water for a couple of minutes to ensure that the temperatures of the gases are the same as that of the water. Twist the plungers of the syringes before taking each reading to ensure that they are not sticking. Continue taking readings until the temperature is about 70 or 80 °C. Plot graphs of volume against temperature for both gases on the same axes. The nitrogen dioxide/dinitrogen tetroxide mixture will expand more than air as the equilibrium responds to the increase in temperature by producing more nitrogen dioxide. If there is time, take further readings as the water cools to check for leaks.

Visual tips

A white background is essential for the 'effect of pressure' part of the demonstration – a clean laboratory coat will do! The colour change is not easy to see and a second syringe of gas to act as a control would be useful. It is probably better to get the class to predict the result before they see the demonstration so that they know what to expect.

THE ROYAL
SOCIETY OF
CHEMISTRY

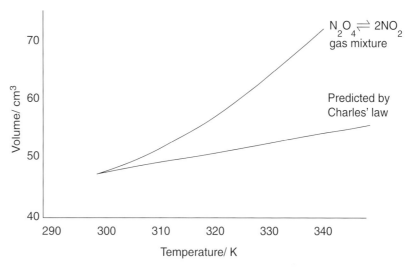

Fig. 3 Comparison of Charles' Law readings with actual results

Teaching tips

The 'effect of temperature' part of the experiment can be done without a second syringe of air if desired. The predicted volume of an ideal gas can be worked out using Charles' Law for each temperature reading and compared with the observed one (*Fig. 3*).

For example if the volume of the gas mixture is 50 cm^3 at 25 °C (298 K), the predicted volume of an ideal gas at 50 °C (323 K) would be 50 x 323/298 = 54.2 cm^3.

The predicted volumes will be less than the observed ones. Students could watch the demonstration with pre-prepared graph axes and plot the points as the readings are taken. They could also do the Charles' Law calculations while watching if necessary.

Theory

The reaction for the preparation of nitrogen dioxide is:

$$Pb(NO_3)_2(s) \rightarrow PbO(s) + 2NO_2(g) + \tfrac{1}{2} O_2(g)$$

The equilibrium is:

$$N_2O_4(g) \rightleftharpoons 2NO_2(g) \qquad \Delta H = +58 \text{ kJ mol}^{-1}$$

Le Chatelier's principle predicts that the equilibrium will move to the right at high temperature and to the left at high pressure.

The equilibrium is almost completely over to the right at 137 °C.

Extensions

The 'effect of temperature' part of the experiment could be interfaced to a suitable computer by using movement sensors (position transducers) to measure the movements of the plungers of both syringes. These could be plotted against time (or, better, against temperature measured with a suitable sensor) 'on-line' and hard copies printed for the students. The book by R. Edwards *Interfacing chemistry experiments.* London: RSC, 1993., gives useful information about computer interfacing.

Further details

It is possible to measure the average relative molecular mass of the gas mixture at any temperature and thus calculate the equilibrium constant if the mass of the gas mixture is found. This can be done by weighing the syringe containing the equilibrium mixture and, later, weighing the same syringe filled with the same volume of air. It is then necessary to allow for the mass of air in the syringe which can be calculated from the density of air at the appropriate temperature (1.293×10^{-3} g cm^{-3} at 273 K).

The relative molecular mass can vary from 46 (pure NO_2) to 92 (pure N_2O_4), a difference of 46. If, for example, the relative molecular mass is found to be 60, then the mixture must contain 14/46 x 100 = 30.4 % N_2O_4 and 32/46 x 100 = 69.6 % NO_2 (*Fig. 4*). If the total pressure is 1 atmosphere, the partial pressures are:

NO_2, 0.696 atm;

N_2O_4, 0.304 atm.

So $Kp = p^2 NO_2(g)_{eqm}/pN_2O_4(g)_{eqm} = 0.696^2/0.304$ atm = 1.59 atm.

Nitrogen dioxide made by the reaction of copper and moderately concentrated nitric acid is not suitable for this experiment unless it is first dried.

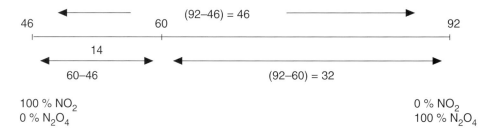

Fig. 4 Equilibrium mixture and relative molecular mass

Acknowledgement

This demonstration has been adapted from one described in *Nuffield advanced science: chemistry – teachers' guide II*. London: Longmans, 1970.

Safety

Wear eye protection.

It is the responsibility of teachers doing this demonstration to carry out an appropriate risk assessment.

THE ROYAL
SOCIETY OF
CHEMISTRY

82. Light scattering by a colloid (the Tyndall effect) – 'the thiosulphate sunset'

Topic

Colloids.

Timing

About 10 min.

Level

Pre-16.

Description

A beam of white light is shone through a solution of sodium thiosulphate and onto a screen. Dilute hydrochloric acid is added to the thiosulphate and colloidal sulphur is formed. The light beam becomes visible in the solution as a bluish 'Tyndall cone' due to light scattered from the colloidal particles. Blue light is scattered more effectively than red so the transmitted light that reaches the screen becomes red and then fades as the colloidal particles become numerous enough to block all the transmitted light. This gives an effect rather like a sunset.

Apparatus

▼ Slide projector with slide holder. (The demonstration can be done using an overhead projector.)

▼ Projection screen. (A white wall will do, or a large sheet of white paper can be used.)

▼ One 1 dm³ beaker.

▼ Access to fume cupboard (optional).

▼ Light meter, for example an Energy Studies Measuring Instrument (ESMI) from British Gas (optional).

Chemicals

The quantities given are for one demonstration.

▼ 20 g of sodium thiosulphate-5-water ($Na_2S_2O_3.5H_2O$)

▼ 1 dm³ of deionised water.

▼ About 1 cm³ of 1 mol dm⁻³ **hydrochloric acid**.

Method

Before the demonstration

Cut a piece of card to the size of a 35 mm slide and make two holes in this each about the diameter of a pencil and as far apart horizontally as possible. Place the card in the slide holder of the projector. This is to produce two beams of light; one will shine through the solution and the other will act as a reference.

If the demonstration is to be done on an OHP, make two holes in a sheet of card, the size of a slide, that covers the OHP stage. These should be positioned about 15 cm apart so that one beam of light shines through a 1 dm³ beaker standing on the stage and the other acts as a reference.

Make up a solution of 20 g of sodium thiosulphate in 1 dm³ of deionised water and place it in the beaker.

The demonstration

Darken the room. Arrange the projector, beaker and screen so that one beam of light shines through the solution in the beaker and onto the screen and the other shines directly onto the screen. Add about 1 cm³ of 1 mol dm⁻³ hydrochloric acid to the thiosulphate solution and stir thoroughly. After about a minute, the solution becomes visibly cloudy and a cone of blue scattered light (the Tyndall cone) becomes visible in the beaker. The transmitted light begins to show on the screen as red. This fades gradually as more sulphur is produced and virtually no light is visible after about five minutes. More concentrated acid will produce the same changes more quickly.

After the experiment, filter the remaining cloudy liquid in a fume cupboard (because of the **sulphur dioxide** which is dissolved in the solution) to show that the filtrate is still cloudy and that the sulphur is colloidal.

Visual tips

At least a partial blackout is desirable. The screen will probably need to be angled towards the audience as shown in *Fig. 1* so that both the transmitted light and the Tyndall cone can be seen.

Teaching tips

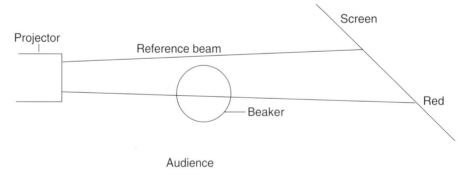

Fig. 1 View from above of apparatus to produce the Tyndall effect

The blue of the sky is caused by blue light scattered by colloidal particles in the atmosphere. The redness of the Sun near sunset is also caused by the scattering of blue light. As the Sun sets, we observe it through an increasing thickness of atmosphere (*Fig. 2*) and more blue light is scattered leaving a higher proportion of red transmitted light.

THE ROYAL
SOCIETY OF
CHEMISTRY

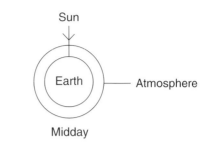

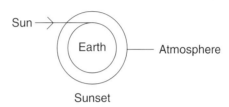

Fig. 2 Daylight and sunset

Theory

The reaction that takes place is:

$$Na_2S_2O_3(aq) + 2HCl(aq) \rightarrow 2NaCl(aq) + SO_2(g) + S(s) + H_2O(l)$$

Extensions

Repeat the experiment holding a light meter in the beam of transmitted light. The reading drops gradually.

Further details

This experiment could usefully be done at the same time as demonstration 78 on following the rate of this reaction using a colorimeter.

Safety

Wear eye protection.

Dispose of the liquid by flushing it down the sink in a fume cupboard with plenty of water.

It is the responsibility of teachers doing this demonstration to carry out an appropriate risk assessment.

THE ROYAL
SOCIETY OF
CHEMISTRY

83. The reaction of hydrogen peroxide and potassium permanganate – 'cannon fire'

Topic

Redox reactions, oxygen as a gas that supports combustion, or general interest.

Timing

A couple of minutes.

Level

Pre-16, but any for general interest.

Description

Potassium permanganate powder is sprinkled onto a burning mixture of hydrogen peroxide solution and ethanol. A series of loud but harmless bangs is heard as the oxygen that is evolved increases the rate of burning.

Apparatus

▼   One evaporating dish, about 10 cm in diameter.

▼   Safety screen.

Chemicals

The quantities given are for one demonstration.

▼   30 cm^3 of 20 volume hydrogen peroxide solution.

▼   20 cm^3 of **ethanol**.

▼   0.5 g of **potassium permanganate** (potassium manganate(VII), $KMnO_4$) powder.

Method

The demonstration

Place the evaporating dish on a large heat-proof mat to protect the bench. Place in the dish 30 cm^3 of 20 volume hydrogen peroxide solution and 20 cm^3 of ethanol and light the mixture with a taper. The ethanol will burn with an almost invisible flame. Now sprinkle about 0.5 g of potassium permanganate into the dish. Immediately, there will be a series of loud bangs giving the effect of a volley of gunshot. This will subside into crackling which will last for up to a minute. This is caused by the reaction of the hydrogen peroxide and potassium permanganate to give oxygen which accelerates the burning of the ethanol. Coarse potassium permanganate powder gives fewer but louder bangs while fine powder gives more but smaller ones. After the reaction is over, brown solid manganese dioxide will be seen in the evaporating dish.

Visual tips

None – this is an audio experiment!

Theory

The reaction between permanganate ions and hydrogen peroxide is:

$$2MnO_4^-(aq) + 3H_2O_2(aq) \rightarrow 2MnO_2(s) + 2H_2O(l) + 3O_2(g) + 2OH^-(aq)$$

Safety

Wear eye protection.

Use a safety screen.

It is not recommended to try other quantities or more concentrated hydrogen peroxide.

It is the responsibility of teachers doing this demonstration to carry out an appropriate risk assessment.

Acknowledgement

This demonstration was suggested by Andrew Szydlo of Highgate School.

84. Zinc-plating copper and the formation of brass – 'turning copper into 'silver' and 'gold''

Topic

Electrochemistry, electroplating, alloys. This is also a fun demonstration for younger children.

Timing

About 10 min.

Level

This is suitable for any age group depending on the sophistication of the theoretical treatment used (if any).

Description

A 'copper' coin is dipped into a solution of sodium zincate in contact with zinc. The coin is plated with zinc and appears silver in colour. The plated coin is held in a Bunsen flame for a few seconds and the zinc and copper form an alloy of brass. The coin now appears gold.

Apparatus

▼   One 250 cm³ beaker.

▼   Bunsen burner, tripod and gauze.

▼   Pair of tongs or tweezers.

▼   Access to top pan balance.

Chemicals

The quantities given are for one demonstration.

▼   5 g of zinc powder.

▼   24 g of **sodium hydroxide** pellets.

▼   A little steel wool or proprietary mild abrasive cleaner.

▼   100 cm³ of deionised water.

▼   Copper coins (*eg* 1 p and 2 p pieces) or copper foil.

Method

Before the demonstration

Dissolve 24 g of sodium hydroxide in 100 cm³ of deionised water. Add 5 g of zinc powder to this solution and heat to boiling point on a Bunsen burner. The solution will fizz as some of the zinc dissolves forming sodium zincate and giving off hydrogen.

Clean a 'copper' coin with steel wool or other mild abrasive cleaner until it is shiny.

THE ROYAL
SOCIETY OF
CHEMISTRY

The demonstration

Drop the cleaned coin into the hot solution containing sodium zincate and the remaining zinc powder. The coin must be in contact with the zinc. Leave the coin until it is plated with a shiny coat of zinc. This will take about 2 – 3 minutes. Leaving the coin too long may cause lumps of zinc to stick to it. Remove the plated coin with tongs or tweezers and rinse it under a running tap to remove any sodium zincate. Show the silver coin to the audience.

Using tongs or tweezers, hold the plated coin in the upper part of a roaring Bunsen flame for a few seconds until the surface turns gold. Turn the coin so that both sides are heated equally. Overheating will cause the coin to tarnish. The gold colour is brass formed by the zinc migrating into the surface layer of the copper. Allow the coin to cool and show it to the audience.

Visual tips

If the mixture of sodium zincate solution and zinc powder is cloudy, filter off the zinc to leave a clear filtrate and place a small piece of zinc foil in the liquid which can then be used for plating.

Teaching tips

Younger students could possibly be told a story about changing copper into silver and gold. It is likely that students will want their own coins plated.

Theory

The formation of sodium zincate is as follows:

$$Zn(s) + 2NaOH(aq) + 2H_2O(l) \rightarrow Na_2Zn(OH)_4(aq) + H_2(g)$$

The plating reaction involves an electrochemical cell; it will not take place unless the copper and the zinc are in contact, either directly or by means of a wire.

The electrode reactions are:

at the zinc electrode:

$$Zn(s) \rightarrow Zn^{2+}(aq) + 2e^-$$

followed by complexing of the zinc ions as $Zn(OH)_4^{2-}(aq)$

at the copper electrode:

$$Zn(OH)_4^{2-}(aq) + 2e^- \rightarrow Zn(s) + 4OH^-(aq)$$

A similar zinc plating process is used industrially but with cyanide ions rather than hydroxide ions as the complexing agent.

Brass is an alloy of copper containing between 18 % and 40 % of zinc.

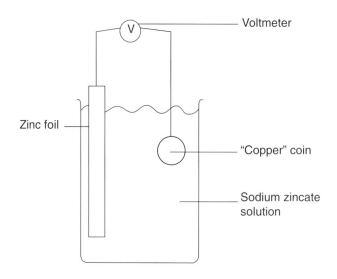

Measuring the emf of the cell produced from copper and zinc

Extensions

Plating of zinc onto copper is, at first sight, unexpected. The reaction could form the basis of a project to investigate factors such as:

▼ The emf of the cell produced (*see figure*);

▼ the effect of the concentration of zincate ions;

▼ the effect of temperature; and

▼ whether other metals can be plated.

Further details

Strictly speaking, it is illegal to 'deface coins of the realm' so the punctiliously law-abiding may prefer to use copper foil or foreign coins. If the latter are used, check beforehand that they work because many different alloys are used in coinage.

Safety

Wear eye protection.

Any remaining finely powdered zinc should not be left to dry because it can ignite spontaneously. Dispose of it by rinsing with water and dissolving in excess dilute sulphuric acid and washing the resulting zinc sulphate solution down the sink.

It is the responsibility of teachers doing this demonstration to carry out an appropriate risk assessment.

THE ROYAL
SOCIETY OF
CHEMISTRY

85. The electrolysis of molten lead bromide

Topic

Electrolysis.

Timing

About 10 min to demonstrate conduction of the molten electrolyte and to show that bromine is produced. About half an hour to collect and weigh the lead that is produced.

Level

Pre-16.

Description

Lead bromide is shown not to conduct in the solid state but does so when molten. The products of the electrolysis are identified as bromine (at the anode) and lead. The lead can be recovered and weighed and a value for the Faraday constant estimated.

Apparatus

▼ Three porcelain crucibles, about 2.5 cm in diameter.

▼ Two lengths of graphite rod, about 15 cm long.

▼ One rubber bung with two holes about 1 cm apart to take the graphite rods.

▼ DC power supply adjustable up to 12–15 V.

▼ Rheostat, if the power supply is not fully adjustable.

▼ 12 V, 5 W bulb, such as a car sidelight bulb and holder.

▼ Ammeter, 0–5 A. A large demonstration model is ideal.

▼ Leads and crocodile clips for the electrical connections.

▼ Stopclock or sight of a clock with a second hand.

▼ Access to a fume cupboard.

▼ Wooden board.

Chemicals

The quantities given are for one demonstration.

▼ About 20 g of **lead bromide** ($PbBr_2$).

▼ A little universal indicator paper.

Method

Before the demonstration

Assemble the apparatus shown in the figure on a wooden board such as a large dissecting board. This allows the audience to see the melting of the lead bromide and the fact that it conducts when molten on the open bench. The apparatus can then be transferred, still assembled on the board, to the fume cupboard for the electrolysis (which produces bromine vapour) to proceed.

THE ROYAL
SOCIETY OF
CHEMISTRY

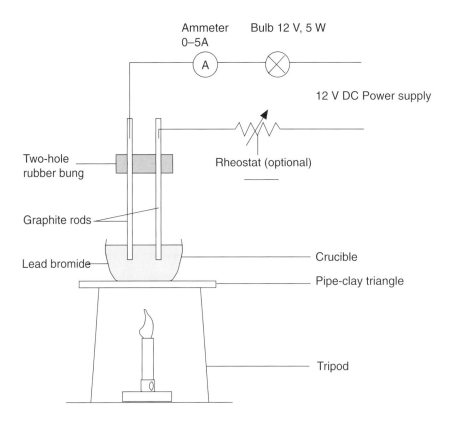

The graphite rods should reach almost to the bottom of the crucible because the volume of the lead bromide powder shrinks considerably in volume on melting. For the same reason, the crucible should be filled to overflowing with lead bromide powder. Slide the two-holed bung as far up the graphite electrodes as possible to prevent it melting or burning with the heat of the Bunsen flame.

Weigh and mark one of the crucibles.

The demonstration

Set the power supply at 12 V and briefly short-circuit the graphite rods to show that the electrical circuit works. Note that the solid lead bromide does not conduct. Light the Bunsen burner and use a medium-sized roaring flame to heat the crucible. The lead bromide melts to a brownish liquid after about five minutes and will begin to conduct electricity. (It may be necessary to add more lead bromide to keep the level of the liquid above the electrodes because the powder shrinks in volume considerably on melting.) The bulb will light and the ammeter will read about 1.5–2 A when all of the solid has melted. Bubbles will be visible at the anode and a brown gas (**bromine**) will be evolved. Hold a piece of moist indicator paper in the gas to show that it is bleached. Turn off the Bunsen burner and allow the lead bromide to solidify; it will stop conducting and bubbles will no longer be seen at the anode. Keep the length of this part of the demonstration to a minimum because bromine vapour is toxic.

Now transfer the apparatus to a fume cupboard. Re-melt the lead bromide and allow the electrolysis to proceed for 15 minutes. Use the power supply and/or rheostat to adjust the current to a convenient value (about 1.5 A) and to maintain the current at this value for the whole 15 minutes.

At the end of this time, switch off the power supply. Using tongs, pick up the crucible and pour the molten lead bromide into a second crucible. A small bead of

THE ROYAL
SOCIETY OF
CHEMISTRY

molten lead will be seen. Retain this in the first crucible and allow it to solidify. When cool, break off any solid lead bromide adhering to the lead, wash and dry the lead and place the bead in the weighed crucible and re-weigh to find the mass of lead. Show that the metal is soft enough to be scratched with a fingernail and pass it round the class. It is possible to write on paper with the lead as with an old fashioned lead (*ie* not graphite) pencil.

Visual tips

During the electrolysis in the fume cupboard, the students can file past the apparatus in small groups to observe the bromine gas evolved at the anode and they can repeat the moist indicator paper test themselves.

Teaching tips

The lead bromide can be re-used.

Most students will readily accept that since the bromine appears at the anode, then the lead is probably discharged at the cathode.

Theory

The reactions are as follows:

at the anode:

$$2Br^-(l) \rightarrow Br_2(g) + 2e^-$$

at the cathode:

$$Pb^{2+}(l) + 2e^- \rightarrow Pb(l)$$

The number of coulombs required to deposit the weighed mass of lead is calculated by multiplying the current in amps by the time in seconds. The value obtained for the Faraday constant will only be approximate due to fluctuations in the current, lead bromide adhering to the lead pellet and the fact that some current will flow during the preliminary part of the demonstration and this will not have been measured (unless this is done as a separate experiment).

Extensions

Some teachers may wish to show that the decomposition is not brought about by heat alone.

Further details

The melting point of lead bromide is unusually low for an ionic compound, 373 °C. It is also possible to melt lithium chloride (T_m 605 °C) with a Bunsen flame in the same apparatus to show that this salt, too, conducts when molten but not when solid.

Safety

Wear eye protection.

It is the responsibility of teachers doing this demonstration to carry out an appropriate risk assessment.

THE ROYAL
SOCIETY OF
CHEMISTRY

86. The liquefaction of chlorine

Topic

Halogens, changes of state.

Timing

About 5 min.

Level

Pre-16 or post-16.

Description

A stream of chlorine from a chlorine generator is directed onto a cold finger condenser containing dry ice and ethanol. Liquid chlorine condenses as yellow drops.

Apparatus

▼   Chlorine generator: one 500 cm³ conical flask with a side arm and a one-holed rubber bung to take a tap funnel, or a 500 cm³ conical flask with a two-holed bung fitted with a tap funnel and a short, right-angled delivery tube. Alternatively, use a chlorine cylinder if available.

▼   One 1 dm³ Buchner flask with a two-holed rubber bung to fit. One hole should be large enough to take a test-tube and the other hole should be fitted with a short length of glass delivery tube.

▼   Short length of rubber tubing to connect the two flasks.

▼   One boiling tube (optional).

▼   Bunsen burner (optional).

▼   One 250 cm³ beaker (optional).

▼   Access to fume cupboard.

Chemicals

The quantities given are for one demonstration.

▼   About 10 g of **potassium permanganate** (potassium manganate(VII), KMnO$_4$).

▼   About 50 cm³ of **concentrated hydrochloric acid**.

▼   About 20 cm³ of **ethanol.**

▼   A few small pellets of dry ice (solid carbon dioxide). This can often be obtained from a local university, hospital or industry. It can be stored for some hours in an expanded polystyrene box or vacuum flask. It can also be made from a carbon dioxide cylinder with a dry ice making attachment.

▼   About 60 g of common salt (sodium chloride) (optional).

▼   About 60 g of crushed ice (optional).

▼   1 cm³ of **bromine** in a sealed ampoule (optional).

▼   A few crystals of solid **iodine** (optional).

THE ROYAL
SOCIETY OF
CHEMISTRY

Method

Before the demonstration
In a fume cupboard, set up the apparatus shown in the figure.

The demonstration
Fill the test-tube about two-thirds full of dry ice chips and add a little ethanol slowly. This will bubble, vigorously at first, as the carbon dioxide sublimes. When the bubbling has settled down, drip the hydrochloric acid slowly on to the potassium permanganate. Green **chlorine** gas is produced and will gradually fill both flasks. After about a minute, yellow drops of liquid chlorine will begin to condense on the cold finger and drop off onto the bottom of the Buchner flask. At first they will vaporise but after a few minutes they will begin to collect as the base of the flask cools down. The base of the flask could be pre-cooled by dry ice or ice/salt mixture if desired.

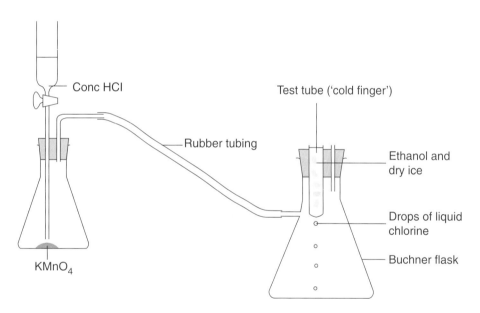

Liquefaction of chlorine

Extensions

This experiment could be combined with a demonstration of the physical states of bromine and of iodine.

Iodine
Heat a few crystals of iodine gently in a boiling tube. The crystals first melt to a dark liquid and then form a purple vapour. This will condense higher up the wall of the boiling tube, reforming solid iodine.

Bromine
Make a mixture of about 60 g of crushed ice and 60 g of salt. Place a sealed ampoule of bromine into this and leave for about five minutes. The bromine will solidify and this can be shown by inverting the ampoule and showing that the bromine remains at the top. At room temperature, the space above the liquid bromine will contain brown bromine vapour.

Teaching tips

Point out the trend in melting and boiling temperatures of the halogens and relate these to van der Waals forces. Ask students to predict the values for fluorine and astatine.

Further details

The relevant data on boiling and melting temperatures are:

	$T_m/\,°C$	$T_b/\,°C$
Chlorine	-101	-35
Bromine	-7	59
Iodine	114	184

The temperature of dry ice is about -80 °C. The salt/ice mixture described attains a temperature of about -15 °C. Some textbooks state that iodine sublimes. Under these conditions, the liquid state can be seen clearly.

One teacher involved in the trials suggested drying the chlorine to prevent any ice condensing on the cold finger.

Safety

Wear eye protection.

After the experiment, allow any remaining liquid chlorine to evaporate before dismantling the apparatus in the fume cupboard.

It is the responsibility of teachers doing this demonstration to carry out an appropriate risk assessment.

THE ROYAL
SOCIETY OF
CHEMISTRY

87. The equilibrium between bismuth oxide chloride and bismuth trichloride

Topic

The effect of concentration on equilibrium position (Le Chatelier's principle).

Timing

About 5 min.

Level

Pre-16 or post-16.

Description

Bismuth oxide chloride is dissolved in concentrated hydrochloric acid to give a clear solution of bismuth trichloride. Addition of water re-forms the bismuth oxide chloride as a white precipitate while subsequent addition of hydrochloric acid re-dissolves this. These changes can be repeated several times in sequence.

Apparatus

▼ Beakers – one each of the following sizes: 25 cm³, 50 cm³, 100 cm³, 250 cm³, 600 cm³ and 2000 cm³. These sizes are not critical, the size larger than that specified could easily be substituted in each case.

▼ Measuring cylinders: four of 10 cm³ and two of 50 cm³ or 100 cm³. It is possible to manage with just one cylinder of each size by measuring the acid out into suitably sized beakers.

▼ Stirring rod.

Chemicals

The quantities given are for one demonstration.

▼ 2.6 g of **bismuth oxide chloride** (bismuth oxychloride, BiOCl).

▼ 83 cm³ of **concentrated hydrochloric acid**.

Method

Before the demonstration
Weigh out 2.6 g (0.01 mol) of bismuth oxide chloride.

Line up on the demonstration bench six beakers of the sizes shown containing water as shown in the table. Beside each beaker place a measuring cylinder (or other suitable container such as a small beaker) containing the volume of concentrated hydrochloric acid shown in the table.

THE ROYAL
SOCIETY OF
CHEMISTRY

Beaker number	Size of beaker / cm³	Volume of water in beaker / cm³	Volume of conc HCl next to beaker / cm³
1	25	12	2
2	50	25	2
3	100	50	3
4	250	125	7
5	600	300	16
6	2000	1000	50

The demonstration

Dissolve the 2.6 g of bismuth oxide chloride in 3 cm³ of concentrated hydrochloric acid in a boiling tube or small beaker. This will give a clear solution of **bismuth trichloride**. Pour this solution into the water in beaker 1. A white precipitate of bismuth oxide chloride will appear immediately as the equilibrium

$$BiOCl(s) + 2HCl(aq) \rightleftharpoons BiCl_3(aq) + H_2O(l)$$

is displaced to the left by the increased concentration of water. Now add the 2 cm³ of hydrochloric acid to beaker 1 and stir. The precipitate re-dissolves as bismuth trichloride as the increased concentration of acid moves the equilibrium to the right.

Now pour the contents of beaker 1 into beaker 2. The precipitate re-appears and can be dissolved again on addition of the pre-measured volume of hydrochloric acid with stirring.

Pour the contents of beaker 2 into beaker 3 followed by addition of acid and so on. Precipitates continue to appear and re-dissolve as predicted by Le Chatelier's principle although the precipitate takes noticeably longer to re-dissolve as the solutions get more dilute.

Visual tips

A dark background is better for viewing the precipitates than a white one.

Teaching tips

Point out that the reaction gets slower as the solutions get more dilute.

Theory

The equilibrium involved is:

$$BiOCl(s) + 2HCl(aq) \rightleftharpoons BiCl_3(aq) + H_2O(l)$$

Further details

This experiment can be started from a solution of bismuth trichloride in water if bismuth oxide chloride is not available.

Safety

Wear eye protection.

It is the responsibility of teachers doing this demonstration to carry out an appropriate risk assessment.

Acknowledgement

This procedure was suggested by Trevor Read of Finchley Catholic High School.

THE ROYAL
SOCIETY OF
CHEMISTRY

88. Catalysts for the thermal decomposition of potassium chlorate

Topic

Reaction rates, catalysis.

Timing

About 5 min to demonstrate catalysis, depending on the number of catalysts demonstrated. It will take about a further 10 min to demonstrate the recovery of the catalyst plus some time to allow the recovered catalyst to dry. The final weighing will probably have to be left until a subsequent lesson.

Level

Pre-16.

Description

Potassium (or sodium) chlorate is heated in a test-tube and the time noted for enough oxygen to be produced to light a glowing taper. The heating is repeated with the addition of various oxide catalysts and the reduced time for the evolution of oxygen is noted. The water insoluble catalyst can be separated from the soluble chlorate salt and weighed to show that it is not used up.

Apparatus

▼ Pyrex test-tubes about 150 mm x 15 mm – one for each catalyst to be shown plus one for the control.

▼ Bunsen burner.

▼ Retort stand with boss and clamp.

▼ Filter funnel.

▼ Conical flask, about 1 dm³ to collect filtrate.

▼ Access to top pan balance and an oven.

▼ Watch glass, a little larger than the filter paper.

▼ Stopwatch or stopclock.

Chemicals

The quantities given are for one demonstration.

▼ 5 g of **potassium chlorate** (potassium chlorate(V), $KClO_3$). This is sufficient to demonstrate one catalyst plus a control. A further 2.5 g will be required for each additional catalyst. **Sodium chlorate** (sodium chlorate(V), $NaClO_3$) may be used instead of potassium chlorate.

▼ 0.25 g of each catalyst to be used. Suitable catalysts include: **copper oxide** (copper(II) oxide), (CuO); manganese dioxide, (MnO_2) ; iron(III) oxide, (Fe_2O_3); silicon dioxide (silica gel, SiO_2).

▼ Wash bottle of deionised water.

▼ Filter paper, eg Whatman no. 1.

THE ROYAL
SOCIETY OF
CHEMISTRY

▼   Wooden tapers.

▼   A little mineral wool (optional).

Method

Before the demonstration

Set up a Bunsen burner and a stand and clamp so that a test-tube can be clamped at about 45° with its base about 5 cm above the burner (so that it will be about 2 cm above the tip of the blue cone of the flame when the Bunsen burner is on) – *Fig. 1*. It is worth doing a preliminary experiment to determine a suitable distance with the burner to be used. Mark the positions of the clamp and burner on the bench so that they can be replaced if disturbed accidentally.

Dry some filter papers in an oven if catalyst recovery is to be attempted.

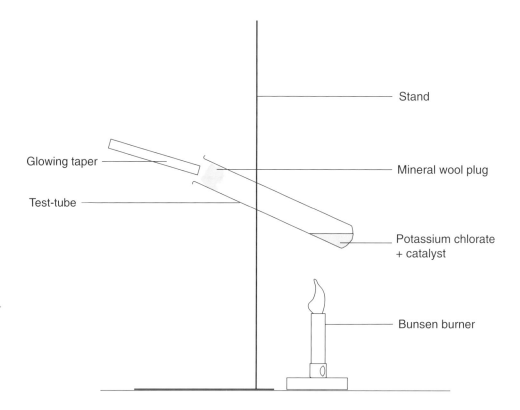

Fig.1 Effect of catalysts on the decomposition of potassium chlorate

The demonstration

Weigh 2.5 g of potassium chlorate into a test-tube and clamp the tube above the burner. Light the burner with gas fully on and air hole open. At the same time start the stopwatch. The solid will melt and begin to give off bubbles of oxygen as it decomposes. Hold a glowing wooden taper in the mouth of the tube until it re-lights. Note the time taken for the taper to relight. This will be about one minute. Turn off the burner.

Now weigh 2.5 g of potassium chlorate into a second identical test-tube, add 0.25 g of copper oxide catalyst and mix well. Clamp the tube as before, light the burner and start the stopwatch. Hold a glowing taper in the mouth of the tube and

THE ROYAL
SOCIETY OF
CHEMISTRY

note the time taken for it to re-light. This will be about one quarter to one third of the time taken without a catalyst.

Repeat the procedure with other catalysts as desired.

It is important that all details are kept the same between runs – the positioning of the tube and burner, the gas flow, the position of the glowing taper *etc.*

Demonstrate the solubility of potassium chlorate and the insolubility of the catalysts in water by shaking a little of each with water in test-tubes. Ask the students to suggest a method of recovering the catalyst to find out whether any of it has been used up.

Catalyst recovery (optional)

Weigh one of the filter papers after they have been dried in the oven. Allow the tube containing the potassium chlorate and copper oxide to cool. Add a little distilled water to the tube and warm gently to dissolve the potassium chlorate. When it has dissolved, filter the contents of the tube through the pre-weighed filter paper, using a wash bottle and distilled water to ensure that all the contents are transferred. Wash the residue on the filter paper two or three times with distilled water to remove any potassium chlorate. Place the filter paper on a watch glass in an oven to dry. When dry (this may need to be the following lesson), re-weigh the filter paper and dry catalyst. With care and good technique, the weight of the recovered catalyst is within a few per cent of the initial amount used.

Visual tips

Catalyst recovery is most obvious with the black catalysts, copper oxide and manganese dioxide.

A large stopclock is ideal.

Teaching tips

There are opportunities to discuss the factors that must be controlled to make the experiment a 'fair test'.

Teachers may prefer not to do the catalyst recovery part of the demonstration, leaving it as a 'thought experiment'.

Theory

The overall reaction for the thermal decomposition is:

$$2KClO_3(l) \rightarrow 2KCl(s) + 3O_2(g)$$

However, it takes place in the following steps:

formation of potassium perchlorate (potassium chlorate(VII))

$$4KClO_3(l) \rightarrow KCl(s) + 3KClO_4(l)$$

decomposition of potassium perchlorate

$$KClO_4(l) \rightarrow KCl(s) + 2O_2(g)$$

It is possible to isolate the perchlorate if the heating is controlled carefully. See, for example, *Nuffield advanced science: chemistry, students' book I*, p123. London: Longman, 1970.

Extensions

Other oxides could be tried as catalysts.
Compare the effectiveness of the different catalysts.
Does the amount of catalyst make a difference?

Further details

An alternative set up would be to clamp two test-tubes above the Bunsen burner so that they are heated equally (*Fig. 2*). The catalyst and control tubes can then be done at the same time. This has a little more impact, but is more difficult to convince the audience that heating of the two test-tubes is equal.

Some teachers may prefer to use equimolar quantities of the catalysts rather than equal masses.

Both sodium chlorate and potassium chlorate react at similar rates.

There is very little difference in effectiveness between the catalysts suggested in this demonstration.

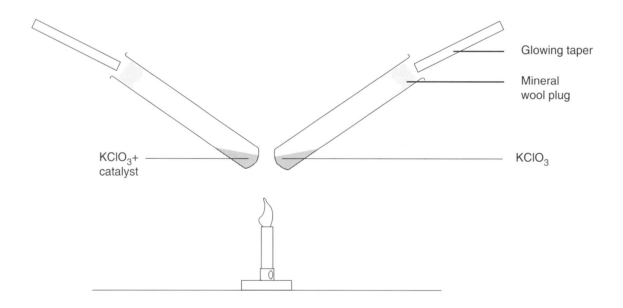

Fig. 2 Alternative method for comparing catalysts

Safety

Wear eye protection.

Avoid dropping the taper into the molten chlorate as this can react quite violently. A loose plug of mineral wool pushed about 1 cm down the test-tube can prevent this. This also ensures that the taper is held at the same position in each experiment.

It is the responsibility of teachers doing this demonstration to carry out an appropriate risk assessment.

THE ROYAL
SOCIETY OF
CHEMISTRY

89. The electrical conduction of silicon – a semiconductor

Topic

Metals, non-metals and semi-metals. Semiconductors.

Timing

About 5 min to demonstrate that the conductivity of silicon increases with temperature whereas that of copper decreases. A further 20 min or so will be needed to plot a graph of current against temperature.

Level

Pre-16 or post-16.

Description

A small piece of silicon is connected in series with a power pack and ammeter. A very small current flows at room temperature, but increases steeply when a hair-drier 'set on hot' is directed at the silicon. Readings of current against temperature can be obtained by immersing the silicon in a heated water bath. The behaviour of silicon can be contrasted with that of a coil of copper wire, where the conductivity decreases with temperature.

Apparatus

▼ Variable voltage power pack, 0 – 12 V.

▼ Voltmeter, dual range, 0 – 5 V and 0 – 15 V.

▼ Ammeter, multi range, 0 – 100 microamps, 0 – 50 mA, and 0 – 1 A. Large demonstration style meters are ideal.

▼ Hair-drier.

▼ Connecting leads and crocodile clips.

▼ Thermometer, 0 – 100 °C.

▼ Bunsen burner, tripod and gauze.

▼ Retort stand, boss and clamp.

▼ One 250 cm³ beaker.

Chemicals

The quantities given are for one demonstration.

▼ One small lump of silicon, 99.999 % pure (available from Aldrich).

▼ About 5 m of enamelled copper wire of approximate diameter 0.2 mm (approximately 36 swg). This will have a resistance of about 2 ohms.

▼ A few lumps of dry ice (optional).

▼ About 50 cm³ of **ethanol** (optional).

▼ Electrically conducting epoxy adhesive (available from RS components).

THE ROYAL
SOCIETY OF
CHEMISTRY

Method

Before the demonstration

Using the conductive epoxy adhesive, glue a short length of insulated connecting wire to each side of a lump of silicon about 10 mm x 5 mm (*Fig. 1*). The adhesive sets after about 15 minutes in a hot oven. If desired, coat the whole assembly in standard (non-conductive) epoxy resin, such as Araldite, to waterproof it. Teachers may prefer to make two assemblies – one to show to students and the other one waterproofed.

   Measure about 5 m of enamelled copper wire of about 0.2 mm diameter and wind it loosely into a coil around a short length of pencil or a small test-tube. Scrape the enamel from a few mm at each end to allow electrical contact.

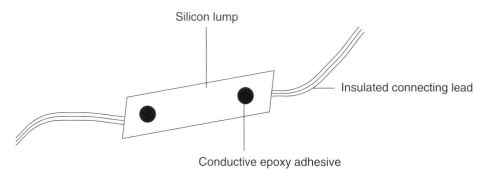

Fig. 1 Conducting silicon lump

The demonstration

Set up the circuit shown in *Fig. 2* using a variable power pack, a micro-ammeter in series with the silicon lump and a 0–15 V voltmeter in parallel with it. With the supply set at 12 V, a current of about 2 microamps flows at room temperature. This will, of course, depend on the temperature and the dimensions of the silicon lump. Hold the hair-drier, set to blow hot air, close to the silicon. The current will rise to over 100 microamps rapidly and drop back again when the hair-drier is removed. The hair-drier will maintain a temperature of around 100 °C depending on its power and how far away from the silicon it is held.

   If desired, plot a current-temperature graph for the silicon as follows:

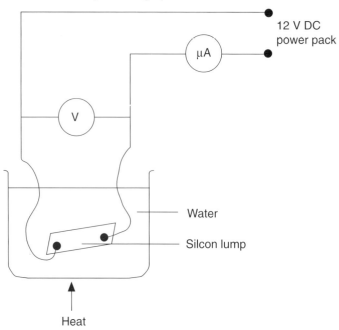

Fig. 2 The electrical conductivity of silicon

THE ROYAL
SOCIETY OF
CHEMISTRY

Place the silicon in a 250 cm³ beaker over a Bunsen burner. Use a waterproofed lump of silicon to ensure that it is clear to the audience that the water cannot be playing any part in the conduction. Waterproofing can be achieved either by coating the silicon in epoxy resin as described above or by placing it in a plastic bag. Heat the water gradually, taking readings of temperature and current every 10° or so. Remove the Bunsen burner and stir the water for a couple of minutes before taking each reading to ensure that the silicon is at the same temperature as the water. At 100 °C the current will be of the order of 300 microamps: the current-temperature graph rises sharply as is typical of a semiconductor.

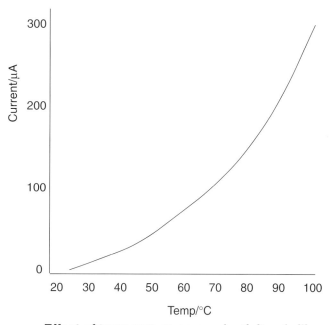

Effect of temperature on conductivity of silicon

If desired, cool the silicon in ice, ice/salt freezing mixture or dry ice/ethanol to take further readings. Take care: the leads have a tendency to unglue themselves in the ethanol/dry ice mixture.

Repeat the experiment with the copper coil in place of the silicon to contrast the behaviour of silicon with that of a metal. This time, set the power pack at 2 V and use a 0 –1 A ammeter. The current will be about 1 A. Note the much better conductivity (lower resistance) of the copper than the silicon, *ie* a much lower potential difference produces a much larger current. Heat the coil with the hair-drier as before and the current will be seen to drop significantly. Dipping the coil in ice, ice/salt mixture or dry ice/ethanol will increase the current. Typical values are 0.62 A under the hair-drier, 0.68 A at room temperature (about 25 °C) and 0.92 A in dry ice/ethanol (about –80 °C).

Visual tips

Large scale demonstration meters are useful. A thermometer that can be connected to a large scale display on a computer monitor is useful. Otherwise students could be asked to take readings and enter them in a table on the blackboard or OHP.

Teaching tips

Pass a piece of silicon around the class. Few students are likely to have seen any despite their familiarity with silicon 'chips'.

THE ROYAL
SOCIETY OF
CHEMISTRY

Theory

A molecular orbital treatment of solids leads to the idea that there are two groups of electronic energy levels or orbitals – the lower called the valence band and the higher called the conduction band *(Fig. 4)*. Electrons in the conduction band are free to move. The energy gap between these levels is responsible for the differences in electrical conductivity between metals (good conductors), non-metals (insulators) and semi-metals (semiconductors). In metals the levels overlap so that there are always electrons in the conduction band to carry a current. This is the 'electron sea'. In insulators there is a large energy gap between the levels so that there are no electrons in the conduction band. Semiconductor bands are close together so that at room temperature a few high energy electrons can jump into the conduction band leaving positively charged 'holes' in the valence band which can also carry current. At higher temperatures, more electrons have enough energy to jump into the conduction band which is why semiconductors conduct better at higher temperatures. In metals, conductivity decreases with temperature because lattice vibrations obstruct the free flow of electrons through the conduction band. The current versus temperature graph for silicon rises sharply because of the statistical distribution of energies among electrons. A small rise in temperature means that a large number of extra electrons will now have enough energy to jump into the conduction band.

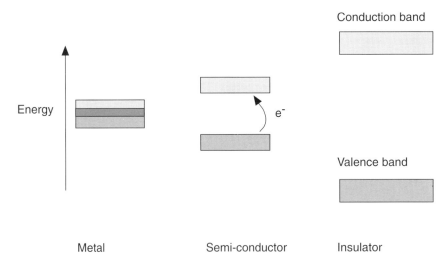

Fig. 4 Conductors, semi-conductors and insulators

Extensions

The resistance of the silicon and of the copper can be calculated at any temperature using 'Ohm's Law', $R = V/I$, where V = potential difference, R = resistance and I = current.

Further details

Do not be tempted to use a wire wound resistor instead of the copper coil. Most commercial resistors are made from alloys that have been designed to have a very small change of resistance with temperature. Unfortunately, crocodile clips do not seem to make satisfactory electrical contact with the silicon and conductive epoxy seems to be necessary.

Safety

Wear eye protection.

It is the responsibility of teachers doing this demonstration to carry out an appropriate risk assessment.

90. Turning 'red wine' into 'water'

Topic

This is a fun demonstration suitable for an open day.

Timing

About 5 min.

Level

Any, for general interest, but a post-16 group could be asked to work out the reactions involved.

Description

A solution of acidified potassium permanganate ('red wine') is poured into a set of glasses previously laced with small amounts of colourless solutions. The 'wine' turns to 'water', 'milk', 'raspberry milk shake' and 'fizzy lemonade'.

Apparatus

▼   One wine (or Ribena) bottle, about 750 cm³ or larger. A laboratory flask will do but the effect is partly lost.

▼   Five wine glasses (or other drinking glasses). Laboratory beakers will do but the effect is partly lost.

Chemicals

The quantities given are for one demonstration.

▼   About 0.4 g of **potassium permanganate** (potassium manganate(VII), $KMnO_4$).

▼   About 24 g of **barium chloride-2-water** ($BaCl_2.2H_2O$).

▼   About 25 g of sodium thiosulphate-5-water ($Na_2S_2O_3.5H_2O$).

▼   About 22 g of sodium carbonate (Na_2CO_3).

▼   About 100 cm³ of 2 mol dm⁻³ **sulphuric acid**.

▼   About 1.5 dm³ of deionised water.

Method

Before the demonstration
Make up the following solutions:

▼   24 g of barium chloride in 100 cm³ of deionised water (approximately 1 mol dm⁻³);

▼   25 g of sodium thiosulphate in 100 cm³ of deionised water (approximately 1 mol dm⁻³);

▼   21 g of sodium carbonate in 100 cm³ of deionised water (approximately 2 mol dm⁻³);

▼   0.4 g of potassium permanganate dissolved in 100 cm³ of 2 mol dm⁻³ sulphuric acid and made up to 1 dm³ with deionised water (approximately 0.0025 mol dm⁻³ with respect to potassium permanganate). Put this solution in a suitable bottle – red wine or Ribena, for example.

THE ROYAL
SOCIETY OF
CHEMISTRY

Line up the five glasses on the bench. Leave the first (glass 1) empty. Then out of sight of the audience, lace the remaining glasses as follows:

▼   glass 2: 1 cm³ of the sodium thiosulphate solution;

▼   glass 3: 1 cm³ of the sodium thiosulphate solution and 3 cm³ of the barium chloride solution;

▼   glass 4: 3 cm³ of the barium chloride solution; and

▼   glass 5: 1 cm³ of the sodium thiosulphate solution and 3 cm³ of the sodium carbonate solution.

The small volumes of liquid will almost certainly not be spotted by the audience. None of the concentrations or volumes is critical.

The demonstration

With a suitable patter, pour a glassful of the permanganate solution into each glass. Do not exceed 200 cm³ for any glass to ensure that the reactants already in the glasses remain in excess.

The following changes will be seen.

▼   Glass 1: no change.

▼   Glass 2: the permanganate will be decolorised as it is reduced to virtually colourless Mn^{2+} ions by the thiosulphate ions. The 'red wine' has turned to 'water' or 'white wine'.

▼   Glass 3: the permanganate will be decolorised as above and a white precipitate of barium sulphate will form as the barium ions react with sulphate ions from the acidified permanganate solution. The 'red wine' has turned to 'milk'.

▼   Glass 4: a white precipitate forms as above, but the colour of the permanganate remains. The 'red wine' has turned to 'raspberry milk shake'.

▼   Glass 5: the permanganate will be decolorised as above and the acidic solution will react with the sodium carbonate to cause bubbles. The 'red wine' has turned into 'fizzy lemonade'.

Visual tips

It is best to remove the 'drinks' fairly quickly as glasses 2 and 5 may gradually go cloudy due to the formation of colloidal sulphur from excess thiosulphate and acid and the white precipitates in glasses 3 and 4 will settle out, spoiling the illusion.

To avoid this, a saturated solution of sodium metabisulphite ($Na_2S_2O_5$) can be used instead of the sodium thiosulphate. However, this has the disadvantage that the initial mixture with barium chloride in glass 3 is cloudy. The cloudiness can be minimised by adding a little dilute hydrochloric acid but this does not remove it entirely and it could be spotted by the audience. This also produces some sulphur dioxide gas which is smelly and may affect asthmatics.

Teaching tips

A good post-16 group could be asked to predict, and write balanced equations for, the reactions used.

THE ROYAL
SOCIETY OF
CHEMISTRY

Theory

The reactions are:

$$2MnO_4^{2-}(aq) + 16H^+(aq) + 10S_2O_3^{2-}(aq) \rightarrow 2Mn^{2+}(aq) + 5S_4O_6^{2-}(aq) + 8H_2O(l)$$

$$Ba^{2+}(aq) + SO_4^{2-}(aq) \rightarrow BaSO_4(s)$$

$$CO_3^{2-}(aq) + 2H^+(aq) \rightarrow H_2O(l) + CO_2(g)$$

Extensions

Other 'drinks' could be devised based on this, or other, systems. For example manipulating the amount of thiosulphate in glass 5 so that the permanganate ends up in slight excess could give 'pink champagne'.

Safety

Wear eye protection.

A warning that chemicals should only be stored in correctly labelled bottles might be appropriate. Do not leave the solutions in the glasses or bottle in case they are mistaken for real drinks.

It is the responsibility of teachers doing this demonstration to carry out an appropriate risk assessment.

THE ROYAL
SOCIETY OF
CHEMISTRY

91. Making rayon

Topic

Polymers.

Timing

The whole demonstration can take up to one and a half hours, much of which is taken up dissolving the cellulose. This can be shortened to 10 or 15 min if a cellulose solution is prepared beforehand.

Level

Pre-16.

Description

Cellulose, in the form of cotton wool or filter paper, is dissolved in a solution containing tetraamminecopper(II) ions to produce a viscous blue liquid. This liquid is injected into sulphuric acid with a hypodermic syringe and fibres of rayon are produced.

Apparatus

▼   Two 250 cm³ beakers.

▼   One 1 dm³ beaker.

▼   Stirring rod.

▼   10 cm³ or 20 cm³ plastic syringe with a hypodermic needle.

▼   Access to fume cupboard.

Chemicals

The quantities given are for one demonstration.

▼   10 g of **basic copper carbonate** (copper(II) carbonate, $CuCO_3.Cu(OH)_2.H_2O$).

▼   100 cm³ of **880 ammonia solution**.

▼   About 500 cm³ of 1 mol dm⁻³ **sulphuric acid**.

▼   2 g of cotton wool (this is about two cotton balls). Check that this is pure cotton and does not contain synthetic fibres. Or 2 g of filter paper (one circle of 18.5 cm diameter is sufficient).

Method

Before the demonstration
Many teachers will wish to prepare a solution of cellulose (see below) before the demonstration to save time. Some may also wish to prepare fibres which have been soaked in acid and washed with water beforehand.

The demonstration
Weigh 10 g of basic copper carbonate into a 250 cm³ beaker and, working in a fume cupboard, add 100 cm³ of 880 ammonia solution. Stir for two minutes and then decant off the resulting deep blue solution, which contains tetraamminecopper(II) ions, into a second beaker.

THE ROYAL
SOCIETY OF
CHEMISTRY

Now add finely shredded cotton wool to the blue solution, slowly and with stirring until the solution has the consistency of shower gel. This will take between 1 and 1.5 g of cotton wool. Alternatively, tear up about 1.5 g of filter paper and use this instead of the cotton wool. Stir until there are no lumps but avoid trapping air bubbles in the liquid. Complete dissolution may take up to an hour. It is probably worth having a pre-prepared solution ready in 'Blue Peter' style to avoid boring the audience.

Take up a few cm³ of this viscous solution, which is called viscose, into a plastic syringe, avoiding taking up any remaining lumps. Fit a hypodermic needle to the syringe and inject a stream of viscose under the surface of about 500 cm³ of 1 mol dm⁻³ sulphuric acid in a 1 dm³ beaker. A thin blue fibre of rayon will be formed. This will slowly turn white as the acid neutralises the alkaline tetraamminecopper(II) solution. After a few minutes, remove the rayon fibre carefully, wash with water and leave to dry on a filter paper. The fibre will be relatively weak.

Theory

Rayon is a so-called regenerated fibre which was once called artificial silk. The polymer contains about 270 glucose units per molecule compared with cotton which contains between 2 000 and 10 000.

Most modern rayons, such as Tenasco and Cordura are produced from a solution called a xanthate which is made by treating cellulose (from wood pulp) with sodium hydroxide and carbon disulphide. The xanthate is then forced through fine holes called spinnerets into an acid bath to re-precipitate the cellulose. Forcing it through a narrow slit produces a sheet of cellophane.

Rayon is used in the manufacture of carpets, tyre cords and surgical materials as well as in clothing.

The experiment described is similar to an older process which produced cuprammonium rayon, known by the trade name of Bemberg.

Extensions

Try other sources of cellulose such as newspaper.

Further details

Glassware can be cleaned with dilute ammonia solution.

Safety

Wear eye protection.

It is the responsibility of teachers doing this demonstration to carry out an appropriate risk assessment.

THE ROYAL
SOCIETY OF
CHEMISTRY

92. The oxidation states of vanadium

Topic

Transition metals – the colours of different oxidation states. Redox reactions and electrode potentials.

Timing

Up to half an hour.

Level

Post-16.

Description

Zinc is used to reduce a yellow solution of ammonium vanadate(V) to a mauve solution containing vanadium(II) ions. The intermediate oxidation states of vanadium(IV) (blue) and vanadium(III) (green) are also seen.

Apparatus

▼   One 1 dm^3 conical flask.

▼   Filter funnel.

▼   Boiling tube.

▼   Dropping pipette.

▼   Four petri dishes (optional).

▼   Access to an overhead projector (optional).

▼   Test-tubes and rack (optional).

Chemicals

The quantities given are for one demonstration.

▼   11.7 g of **ammonium metavanadate** (ammonium vanadate(V), NH_4VO_3).

▼   15 g of zinc powder.

▼   100 cm^3 of approximately 0.25 mol dm^{-3} **potassium permanganate** (potassium manganate(VII), $KMnO_4$) in 1 mol dm^{-3} sulphuric acid. Dissolve 4 g of potassium permanganate in 100 cm^3 of 1 mol dm^{-3} sulphuric acid.

▼   About 1 g of powdered tin (optional).

▼   About 10 cm^3 of approximately 1 mol dm^{-3} sodium thiosulphate solution (optional). Dissolve about 25 g of sodium thiosulphate-5-water in 100 cm^3 of water.

▼   1 dm^3 of 1 mol dm^{-3} **sulphuric acid**.

Method

Before the demonstration

Make up a 0.1 mol dm^{-3} solution of ammonium metavanadate by dissolving 11.7 g of solid in 900 cm^3 of 1 mol dm^{-3} sulphuric acid and making up to 1 dm^3 with deionised water. This yellow solution is usually represented as containing VO_2^+(aq) ions (dioxovanadium(V) ions) in which vanadium has an oxidation number of +5.

THE ROYAL
SOCIETY OF
CHEMISTRY

The demonstration

Place 500 cm^3 of the ammonium metavanadate solution in a 1 dm^3 conical flask and add about 15 g of powdered zinc. This will effervesce and give off hydrogen on reaction with the acid.

The solution will immediately start to go green and within a few seconds will turn pale blue, the colour of the VO^{2+}(aq) ion in which the vanadium has an oxidation number of +4. The short-lived green colour is a mixture of the yellow of V^V and the blue of V^{IV}. The blue colour of VO^{2+} is similar to that of the Cu^{2+}(aq) ion. Over a further fifteen minutes or so, the colour of the solution changes first to the green of V^{3+}(aq) ions and eventually to the mauve of V^{2+}(aq) ions. The green of V^{3+}(aq) is the most difficult to distinguish.

If desired, decant off a little of the solution at each colour stage, filter it to remove zinc and stop the reaction and place in a petri dish on the overhead projector to show the colour more clearly.

When the reaction has reached the mauve stage, filter off a little of the solution into a boiling tube and add acidified potassium permanganate solution dropwise. This will re-oxidise the vanadium through the +3 and +4 oxidation states back to V^V. Take care with the final few drops to avoid masking the yellow colour of vanadium(V) with the purple of permanganate ions.

Visual tips

A white background is vital if the colour changes are to be clearly seen.

If desired, prepare solutions containing V^{IV} and V^{III} beforehand for comparison. This is recommended for teachers who are not familiar with these colours. This can be done as follows.

V^{IV}: take a little of the original ammonium metavanadate solution in a test-tube and add approximately 1 mol dm^{-3} sodium thiosulphate solution dropwise until no further colour change occurs and a light blue solution is obtained. If too much thiosulphate is added, the solution will gradually go cloudy due to the formation of colloidal sulphur by reaction of the excess thiosulphate with acid but this will not affect the blue colour of V^{IV}.

V^{III}: take a little of the original ammonium metavanadate solution in a test-tube and add a spatula-full of powdered tin. Leave this for about five minutes and then filter off the tin to leave a green solution containing V^{3+}(aq) ions.

Teaching tips

This demonstration can be used as an introduction to the idea that different oxidation states of transition metal ions often have different colours and that electrode potentials can be used to help predict the course of redox reactions (via the 'anti-clockwise rule' or otherwise). While waiting for the reaction to go to completion, some of the reactions can be discussed.

THE ROYAL
SOCIETY OF
CHEMISTRY

Theory

The relevant half reactions and redox potentials are as follows:

$$Zn^{2+}(aq) + 2e^- \rightleftharpoons Zn(s) \qquad\qquad E^\ominus = -0.76 \text{ V}$$

$$V^{3+}(aq) + e^- \rightleftharpoons V^{2+}(aq) \qquad\qquad E^\ominus = -0.26 \text{ V}$$

$$Sn^{2+}(aq) + 2e^- \rightleftharpoons Sn(s) \qquad\qquad E^\ominus = -0.14 \text{ V}$$

$$VO^{2+}(aq) + 2H^+(aq) + e^- \rightleftharpoons H_2O(l) + V^{3+}(aq) \qquad E^\ominus = +0.34 \text{ V}$$

$$S_4O_6^{2-}(aq) + 2e^- \rightleftharpoons 2S_2O_3^{2-}(aq) \qquad E^\ominus = +0.47 \text{ V}$$

$$VO_2^+(aq) + 2H^+(aq) + e^- \rightleftharpoons H_2O(l) + VO^{2+}(aq) \qquad E^\ominus = +1.00 \text{ V}$$

So zinc will reduce $VO_2^+(aq)$ to $V^{2+}(aq)$, tin will reduce $VO_2^+(aq)$ to $V^{3+}(aq)$ and no further and thiosulphate ions will reduce $VO_2^+(aq)$ to $VO^{2+}(aq)$ and no further.

Further details

This demonstration would be a good introduction to the experiments involving the redox reactions of vanadium described in *Revised nuffield advanced science, chemistry: students' book II*, p 224. London: Longman, 1984, along with the relevant section in the Teachers' Guide (p1423) and by D. J. Redshaw in *Sch. Sci. Rev.*, 1974, **193**, 753.

Safety

Wear eye protection.
    It is the responsibility of teachers doing this demonstration to carry out an appropriate risk assessment.

THE ROYAL
SOCIETY OF
CHEMISTRY

93. Complexes of nickel(II) with ethylenediamine

Topic

Transition metal chemistry, complex ions.

Timing

About five min.

Level

Post-16.

Description

The ions $[Ni(H_2O)_6]^{2+}$(aq), $[Ni(H_2O)_4(en)]^{2+}$(aq), $[Ni(H_2O)_2(en)_2]^{2+}$(aq) and $[Ni(en)_3]^{2+}$(aq) are produced and their colours are seen to be distinctly different.

Apparatus

▼   Seven 250 cm³ beakers.

▼   One 0 –100 cm³ measuring cylinder.

▼   Dropping pipette.

▼   Access to an overhead projector (optional).

Chemicals

The quantities given are for one demonstration.

▼   9.52 g of **nickel(II) chloride-6-water** ($NiCl_2.6H_2O$).

▼   12 g of **ethylenediamine** (1,2-diaminoethane, $NH_2CH_2CH_2NH_2$).

▼   About 20 cm³ of **concentrated hydrochloric acid**.

▼   About 1.5 dm³ of deionised water.

Method

Before the demonstration
Make up a 0.2 mol dm⁻³ solution of nickel(II) chloride by dissolving 9.52 g of nickel(II) chloride-6-water in deionised water and making it up to 200 cm³.

Make up a 0.2 mol dm⁻³ solution of ethylenediamine by dissolving 12 g of ethylenediamine in deionised water and making it up to 1 dm³.

The demonstration
Pour 50 cm³ of the green nickel(II) solution into each of four 250 cm³ beakers. To the first beaker add 50 cm³ of the ethylenediamine solution. This will turn pale blue as the tetraaqua(1,2-diaminoethane)nickel(II), $[Ni(H_2O)_4(en)]^{2+}$(aq), ion is formed.

To the second beaker add 100 cm³ of the ethylenediamine solution. This will turn blue-purple as the diaquabis(1,2-diaminoethane)nickel(II), $[Ni(H_2O)_2(en)_2]^{2+}$(aq), ion is formed.

To the third beaker add 150 cm³ of the ethylenediamine solution. This will turn violet as the tris(1,2-diaminoethane)nickel(II), $[Ni(en)_3]^{2+}$(aq), ion is formed.

Take each of the solutions of the complexes and pour about half into an empty beaker. Add concentrated hydrochloric acid dropwise to each solution. The colour changes will be reversed back to the green hexaaqua ion. This is because the ligands exchange fairly easily with water molecules and the acid protonates their lone pairs, leaving them unable to co-ordinate with the metal ion.

Visual tips

A white background is important for the colour changes to be seen clearly. Alternatively, the beakers can be placed on an overhead projector to show the colours.

Teaching tips

Point out that ethylenediamine is being added to the nickel(II) ions in 1:1, 2:1 and 3:1 molar ratios respectively.

The point that the protonated ethylenediamine is no longer a ligand is worth stressing.

Theory

The series of reactions is:

$$[Ni(H_2O)_6]^{2+}(aq) + en(aq) \rightarrow [Ni(H_2O)_4(en)]^{2+}(aq) + 2H_2O(l)$$

$$[Ni(H_2O)_4(en)]^{2+}(aq) + en(aq) \rightarrow [Ni(H_2O)_2(en)_2]^{2+}(aq) + 2H_2O(l)$$

$$[Ni(H_2O)_2(en)_2]^{2+}(aq) + en(aq) \rightarrow [Ni(en)_3]^{2+}(aq) + 2H_2O(l)$$

Further details

Ethylenediamine solutions are not stable for any length of time as, being strongly basic, they pick up atmospheric carbon dioxide. Make up the solution freshly before each demonstration.

Safety

Wear eye protection.

It is the responsibility of teachers doing this demonstration to carry out an appropriate risk assessment.

THE ROYAL
SOCIETY OF
CHEMISTRY

94. The lead-acid accumulator

Topic

Electrochemistry, everyday chemistry.

Timing

Ten min upwards.

Level

Pre-16.

Description

Lead foil electrodes are placed in a solution of sulphuric acid and connected to a power pack. After charging for a minute or so, the electrodes are connected to a torch bulb which lights for several seconds.

Apparatus

▼ DC power pack, variable from 0 –12 V.

▼ Connecting leads and crocodile clips.

▼ Torch bulb, 2.5 V, 0.25 A, in a suitable holder.

▼ Voltmeter, 0 –5 V (a large demonstration type is ideal).

▼ One 250 cm³ beaker.

▼ Stopwatch or clock (optional).

Chemicals

The quantities given are for one demonstration.

▼ Two pieces of lead foil, each about 4 cm x 8 cm.

▼ About 200 cm³ of 1 mol dm⁻³ **sulphuric acid**.

▼ A little **lead(IV) oxide** (lead dioxide, PbO_2) (optional).

Method

The demonstration

Place the lead foil strips down either side of a 250 cm³ beaker and clip them in place with crocodile clips. Place 200 cm³ of 1 mol dm⁻³ sulphuric acid in the beaker. Connect the lead electrodes to the power pack set at 2 V and switch on for about two minutes. Point out that the lead connected to the positive terminal becomes covered with chocolate brown lead dioxide. If desired, remove the electrode from the beaker to show this to the audience and compare the colour with that of a sample of known lead dioxide to confirm its identity.

Disconnect the leads from the power pack and connect them to the torch bulb. This will glow for a few seconds indicating that the cell has stored some electricity. Note that the brown colour on the positive electrode does not disappear on discharge. Measure the potential difference between the electrodes on discharge. This will be 2 V.

There are now various possibilities, for example:

1.   Demonstrate the effect of charging the battery for different lengths of time, increasing the charging time in steps of, say, five seconds. The bulb remains lit for longer, up to a charging time of about 30 seconds after which little difference will be observed.

2.   Demonstrate the effect of charging the battery at different potential differences. The charging voltage makes little difference to the time for which the bulb will light. Potential differences greater than 2 V produce gases at the electrodes. It is possible to collect these in an inverted test-tube and show that they are hydrogen at the negative electrode and oxygen at the positive.

Other possibilities include investigating the effect of surface area of the electrodes, the effect of other electrolytes (hydrochloric acid, for example, does not work), how long the battery will hold its charge for *etc.*

Teaching tips

This demonstration can be used to introduce a wide variety of student investigations based on this system as there is a large number of variables to investigate.

Theory

The initial reactions form a coating of lead dioxide on the positive electrode as follows.

At the positive electrode:

$$Pb(s) + 2H_2O(l) \rightarrow PbO_2(s) + 4H^+(aq) + 4e^-$$

At the negative electrode:

$$2H^+(aq) + 2e^- \rightarrow H_2(g)$$

Some white, insoluble lead sulphate is also formed by reaction of the lead with sulphuric acid.

Subsequently, the following reactions occur on charging and are reversed on discharge.

At the positive electrode (the one coated with lead dioxide):

$$PbSO_4(s) + 2H_2O(l) \rightarrow PbO_2(s) + 4H^+(aq) + SO_4^{2-}(aq) + 2e^-$$

At the negative electrode:

$$PbSO_4(s) + 2e^- \rightarrow Pb(s) + SO_4^{2-}(aq)$$

The overall effect is the production of sulphuric acid on charging and its consumption on discharge. Hence the state of charge can be monitored by measuring the density (or specific gravity) of the electrolyte with a hydrometer as is done in garages. Sulphuric acid has almost twice the density of water.

In theory, the cell reactions are totally reversible but in practice material gradually falls off the electrodes.

The potential difference of the cell is just over 2 V which falls slightly on discharge. If the cell is charged at a potential difference much greater that 2 V, water is electrolysed giving rise to hydrogen bubbles at the negative electrode and oxygen at the positive electrode.

Extensions

An old car battery could be drained of its acid and cut open to show its construction although great care should be taken to ensure that all the acid has been removed.

The total amount of charge stored by the battery could be estimated by discharging it into potassium iodide solution and titrating the liberated iodine with thiosulphate.

On discharging the battery into potassium iodide solution, iodine is produced at the positive electrode:

$$2I^-(aq) \rightarrow I_2(aq) + 2e^-$$

This can be titrated with standard sodium thiosulphate solution in the usual way using starch at the end point.

$$I_2(aq) + 2Na_2S_2O_3(aq) \rightarrow 2NaI(aq) + 2Na_2S_4O_6(aq)$$

So each mole of thiosulphate ions is equivalent to a mole of electrons stored in the battery.

Further details

Car batteries are lead-acid accumulators containing six cells connected in series to provide a potential difference of 12 V.

Mechanics are warned against smoking when charging batteries because of the hydrogen which is evolved.

Strictly, the demonstration apparatus is a cell and two or more connected together form a battery.

Safety

Wear eye protection.

It is the responsibility of teachers doing this demonstration to carry out an appropriate risk assessment.

95. Making polysulphide rubber

Topic

Polymers.

Timing

About half an hour.

Level

Pre-16 or post-16.

Description

A solution of sodium polysulphide is made by dissolving sulphur in hot sodium hydroxide. This solution is reacted with 1, 2-dichloroethane and a condensation polymer, polysulphide rubber is formed.

Apparatus

▼   Two 400 cm³ beakers.

▼   Bunsen burner, tripod and gauze.

▼   Stirring rod.

▼   Thermometer, 0 –100 °C.

▼   Tongs or tweezers.

▼   One 100 cm³ measuring cylinder.

▼   Access to a fume cupboard.

Chemicals

The quantities given are for one demonstration.

▼   8 g of flowers of sulphur.

▼   5 g of **sodium hydroxide**.

▼   20 cm³ of **1, 2-dichloroethane** (ethylene dichloride, $ClCH_2CH_2Cl$).

▼   A little washing up liquid.

Method

Before the demonstration
Make a solution of 5 g of sodium hydroxide in 100 cm³ of water.

The demonstration
Work in a fume cupboard. Heat the sodium hydroxide solution in the 400 cm³ beaker until it is boiling. Add the 8 g of sulphur and a drop of washing up liquid to help the sulphur mix with the aqueous solution. Continue to boil the mixture and stir until the sulphur has dissolved to give a dark red-brown solution of sodium polysulphide, NaS_xNa, where x is between 2 and 6 and typically 4. This will take about 15 minutes.

Decant the liquid off any small lumps of sulphur which remain undissolved. Allow the liquid to cool to below 83 °C, the boiling point of 1, 2-dichloroethane, and add 20 cm³ of 1, 2-dichloroethane with stirring. Continue to stir for a few minutes as

THE ROYAL
SOCIETY OF
CHEMISTRY

the red-brown liquid turns orange and a lump of whitish rubber about the size of a typical pencil eraser forms in the liquid.

Remove the yellowish-white lump of rubber with tweezers, wash it several times with water and leave in the fume cupboard for a few minutes to allow excess 1, 2-dichloroethane to evaporate. Handling the plastic with gloves, demonstrate its flexibility. Do not pass the rubber round the audience as it tends to be rather smelly due to sulphur compounds and it may still contain traces of 1, 2-dichloroethane.

Theory

Sodium polysulphide is NaS_xNa where x is between 2 and 6. The sulphur chains are linear. The reaction with 1, 2-dichloroethane is:

$$nClCH_2CH_2Cl + nNaS_xNa \rightarrow [CH_2CH_2S_x]_n + 2nNaCl$$

Further details

This type of rubber is called Thiokol A and was one of the earliest synthetic rubbers. Thiokol rubbers have good resistance to solvents and are used to make printing rollers and sealants.

Other organic dihalides can be used instead of 1, 2-dichloroethane to form other Thiokols. Di-2-chloroethyl ether is used to make Thiokol B.

Safety

Wear eye protection.

It is the responsibility of teachers doing this demonstration to carry out an appropriate risk assessment.

96. A hydrogen-oxygen fuel cell

Topic

Electrochemistry, energy in chemistry.

Timing

About ten min.

Level

Pre-16 or post-16.

Description

Two methods are described. In method 1, two platinum electrodes are immersed in sodium hydroxide solution. Hydrogen is bubbled past one and oxygen past the other. A potential difference of almost 1 V is produced across the two electrodes.

In method 2, sodium hydroxide is electrolysed between graphite electrodes and the products (hydrogen and oxygen) are retained around the electrodes. A voltmeter is then connected across the electrodes and registers a potential difference of over 1 V.

Apparatus

Method 1

▼   Two hydrogen electrodes.

▼   One 400 cm³ beaker.

▼   Access to cylinders of hydrogen and oxygen with appropriate valve gear.

▼   One high resistance voltmeter (0–1 V). A pH meter with a scale calibrated in volts is ideal.

▼   Connecting leads and crocodile clips.

Method 2

▼   One electrolysis cell. This is made from a short length of glass or plastic tube of diameter about 3–4 cm fitted with a two-holed rubber bung. A graphite electrode is fitted into each hole in the bung *(Fig. 2)*.

▼   Two test-tubes, about 50 mm x 10 mm. These should be narrow enough so that they can be filled with water and inverted carefully without the water running out. The water is held in by surface tension.

▼   DC power pack, adjustable 0 – 4 V (or more).

▼   One high resistance voltmeter (0 – 1 V). A pH meter with a scale calibrated in volts is ideal.

▼   One wooden taper.

▼   One small rubber band.

▼   Connecting leads and crocodile clips.

Chemicals

The quantities given are for one demonstration.

▼ About 250 cm³ of 1 mol dm⁻³ **sodium hydroxide**.

Method

The demonstration
Method 1

Clamp the two hydrogen electrodes so that their ends are immersed in sodium hydroxide solution in the beaker (*Fig. 2*). Connect one to the **hydrogen** cylinder and the other to the oxygen cylinder via suitable regulators and adjust the gas flows to produce about one bubble per second. Connect the voltmeter across the two electrodes so that the oxygen electrode is connected to the positive terminal. A potential difference of about 0.9 V will be registered.

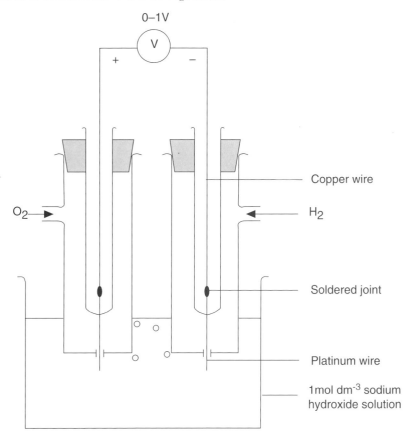

Fig. 1 Apparatus for method 1

Method 2

Clamp the electrolysis cell vertically and fill it with sodium hydroxide solution to just above the level of the electrodes. Fill one of the test-tubes with water and invert it carefully. The water should stay in the tube, held by surface tension. This is worth practising before the demonstration! Carefully lower the tube over one of the electrodes. Repeat with the second tube and place it over the other electrode. Lift the two tubes to expose a few mm of the electrodes and slide a short length of wooden taper between them. Bind the two tubes together with a rubber band and the taper will hold the tubes in position as it rests on top of the cell.

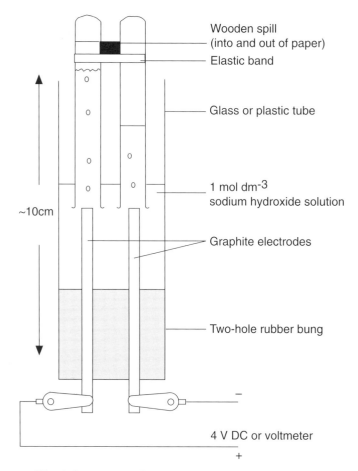

Wooden spill
(into and out of paper)

Elastic band

Glass or plastic tube

1 mol dm^{-3}
sodium hydroxide solution

Graphite electrodes

Two-hole rubber bung

4 V DC or voltmeter

~10cm

Fig. 2 Apparatus for method 2

Connect the electrodes to a power pack set at about 4 V DC and electrolyse the solution. Oxygen is produced at the anode and hydrogen at the cathode. Continue until both test-tubes are full of gas. Now disconnect the power pack and connect a voltmeter across the electrodes with the positive terminal connected to the oxygen electrode. A reading of about 1.2 V will be obtained.

Note, the electrolysis of water should produce twice the volume of hydrogen at the cathode as oxygen at the anode. Significantly less oxygen than this is observed. This is presumably due to reaction of some of the oxygen with the electrode to give carbon dioxide which dissolves in the electrolyte.

Visual tips

Take care when setting out the apparatus, especially with method 1 where tubes and leads can easily obscure the beaker and electrodes.

Teaching tips

Method 1 is the more satisfactory if hydrogen electrodes and gas cylinders are available. If using method 2, it will be necessary to explain the difference between a storage battery and a fuel cell and explain that the electrolysis stage is just a convenient way of obtaining the fuel gases.

THE ROYAL
SOCIETY OF
CHEMISTRY

Theory

The reactions are:

at the oxygen electrode:

$$O_2(g) + 2H_2O(l) + 4e^- \rightarrow 4OH^-(aq) \qquad E^\circ = +0.40 \text{ V}$$

at the hydrogen electrode:

$$2H_2(g) + 4OH^-(aq) \rightarrow 4H_2O(l) + 4e^- \qquad E^\circ = +0.83 \text{ V}$$

This gives an overall reaction of:

$$2H_2(g) + O_2(g) \rightarrow 2H_2O(l) \qquad E^\circ = +1.23 \text{ V}$$

The expected potential difference of 1.23 V is not achieved in practice because of other equilibria involved.

Extensions

The fact that the hydrogen/oxygen can release energy in other forms can be shown by a suitable demonstration of an explosion or burning hydrogen at a jet. See, for example, demonstrations numbers 35, 36, 37 and 66.

Further details

In method 1, if a lower voltage than about 0.9 V is obtained, it may be necessary to re-platinise the electrodes. This can be done by following the method given in *The CLEAPSS school science service laboratory handbook*, p1728. Brunel: CLEAPSS, 1992.

If available, a small electrical device that will operate at the potential difference produced by the cell would be more effective than a voltmeter in making the point that energy is produced. Suitable devices might be a small electric motor or a light emitting diode (LED).

Safety

Wear eye protection.

It is the responsibility of teachers doing this demonstration to carry out an appropriate risk assessment.

THE ROYAL
SOCIETY OF
CHEMISTRY

97. Light sensitive silver salts

Topic

Photochemistry, everyday chemistry (photography).

Timing

About 15 min.

Level

Pre-16.

Description

A precipitate of silver chloride, made by the reaction of silver nitrate and sodium chloride, is filtered at the pump. Opaque masks such as coins are placed on the filter paper which is then placed under a photoflood bulb. The silver chloride that is exposed to the light reacts to form silver and turns grey while that under the masks remains white. Silver bromide works similarly.

Apparatus

▼ One 400 cm³ beaker.

▼ Buchner funnel (about 7 cm diameter or larger).

▼ Buchner flask (500 cm³ or 1 dm³), and adapter.

▼ Access to filter pump.

▼ Filter paper such as Whatman no. 40 to fit the Buchner funnel.

▼ 275 W photoflood bulb with suitable holder.

▼ Coins or other small opaque objects to act as masks.

▼ Pair of scissors.

▼ Dropping pipette.

Chemicals

The quantities given are for one demonstration.

▼ 500 cm³ of 0.01 mol dm⁻³ **silver nitrate**.

▼ 250 cm³ of 0.1 mol dm⁻³ sodium chloride.

▼ 250 cm³ of 0.1 mol dm⁻³ potassium bromide.

▼ 50 cm³ of 1 mol dm⁻³ sodium thiosulphate.

Method

Before the demonstration
Set up the suction filtration apparatus.

The demonstration
Mix 100 cm³ of silver nitrate solution with 50 cm³ of sodium chloride solution. Filter the resulting white precipitate of silver chloride at the pump. If the ambient light is bright, cover the funnel with aluminium foil to prevent premature exposure of the

THE ROYAL
SOCIETY OF
CHEMISTRY

silver chloride. It is not important if some silver chloride comes through the filter paper. Remove the silver chloride-impregnated filter paper from the funnel and place it on a larger filter paper or piece of blotting paper on the bench. Place a photoflood bulb in a suitable holder about 10 cm away from the silver chloride-impregnated filter paper. Place two coins (or other small opaque objects) on the filter paper to mask the light and switch the bulb on. Within about two minutes the un-masked paper will turn grey due to the presence of silver. Switch off the bulb and remove the coin masks to show that the silver chloride below them remains unchanged white.

Cut the filter paper in half so that one of the unexposed circles is on each half. Take one half and, using a dropping pipette, soak the unexposed circle several times with sodium thiosulphate solution, allowing the solution to drip into the sink. Use about 5 cm³ of solution in total. Rinse the filter paper with water from a wash bottle to remove thiosulphate. This treatment will dissolve away most of the unchanged silver chloride from the unexposed circle and mimics the fixing stage of the photographic process.

Replace both halves of the filter paper under the photoflood bulb without the coin masks and switch on for a further two minutes or so. The untreated circle will turn grey, so that it is virtually the same colour as the previously exposed part of the filter paper. The circle that has been treated with sodium thiosulphate will remain mostly white.

Visual tips

If the filter paper could be mounted vertically by pinning to a cork pin board, the audience could see the colour change more clearly. The masks could then be, say, drawing pins.

Other objects, such as shapes or initials cut from aluminium foil, could be used as masks.

Teaching tips

Explain that these reactions are the basis of photography.

Explain that photochemical reactions like these are the reason for silver compounds being stored in brown bottles.

Theory

The silver ions in the silver chloride are reduced by the action of light to metallic silver.

$$AgCl(s) \rightarrow Ag(s) + \tfrac{1}{2}\,Cl_2(g)$$

In real photography, a latent image is first formed which is then developed by the use of an organic reducing agent. This is followed by 'fixing' in which unreacted silver halide is dissolved away in sodium thiosulphate solution (called hypo by photographers) as complex ions so that the image can be exposed to light without further blackening. This produces a negative image which is blackest where most light has fallen.

$$AgCl(s) + 2Na_2S_2O_3(aq) \rightarrow Na_3Ag(S_2O_3)_2(aq) + NaCl(aq)$$

Extensions

Repeat using potassium bromide solution instead of sodium chloride. The precipitate of silver bromide is yellowish and reacts faster to the light than the chloride. It is also easier for the demonstrator to see when the yellowish precipitate has dissolved when treating with thiosulphate solution.

Further details

Silver iodide does not appear to be effective for this demonstration, giving no colour change after several minutes of exposure.

Sunlight could be used instead of a photoflood lamp but it is slower and less reliable.

Safety

Wear eye protection.

It is the responsibility of teachers doing this demonstration to carry out an appropriate risk assessment.

THE ROYAL
SOCIETY OF
CHEMISTRY

98. Cracking a hydrocarbon/ dehydrating ethanol

Topic

Petrochemicals/crude oil/industrial chemistry/hydrocarbons, alcohols.

Timing

About ten min.

Level

Pre-16.

Description

Liquid paraffin (a mixture of alkanes of chain length C_{20} and greater) is vaporised and passed over a hot pumice stone catalyst. A gaseous product is obtained which is flammable and which will decolorise bromine water and acidified permanganate ions. The same apparatus and method can be used to dehydrate ethanol.

Apparatus

▼ One boiling tube with a one-holed rubber bung fitted with a delivery tube.

▼ Six test-tubes about 120 mm x 16 mm with corks or bungs to fit.

▼ Bunsen burner.

▼ Test-tube rack.

▼ Glass pneumatic trough, plastic fish tank or washing up bowl.

▼ Dropping pipette.

▼ Bunsen valve (optional). This consists of about 3 cm of rubber tube fitted onto the end of the delivery tube and closed with a short length of glass rod. The tube has a slit cut along its length with a scalpel. This valve may help to prevent suckback of water up the delivery tube but it is not always effective *(Fig.1)*.

▼ Safety screen.

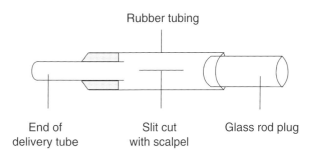

Rubber tubing

End of   Slit cut   Glass rod plug
delivery tube   with scalpel

Fig. 1 A Bunsen valve

Chemicals

The quantities given are for one demonstration.

▼   About 2 cm³ of medicinal liquid paraffin (a mixture of alkanes of chain length C_{20} and greater) or **ethanol**.

▼   A little mineral wool.

▼   About half a boiling tube full of small lumps of pumice stone or broken pieces of porous (*ie* unglazed) pot in lumps about the size of a pea.

▼   A few cm³ of **bromine** water – make sure that this is pale brown in colour.

▼   A few cm³ of a solution which is approximately 0.01 mol dm⁻³ in **potassium permanganate** (potassium manganate(VII)) and approximately 0.1 mol dm⁻³ in **sulphuric acid**. This is achieved by dissolving the potassium permanganate in 0.1 mol dm⁻³ sulphuric acid.

Method

For cracking the hydrocarbon

Before the demonstration

Half fill the pneumatic trough with water. Fill the test-tubes with water and leave them in the trough.

The demonstration

Place a small tuft of mineral wool in the bottom of the boiling tube so that it fills the bottom cm or so. Using a dropping pipette, squirt about 2 cm³ of liquid paraffin into the mineral wool so that it soaks in. It should be possible to invert the tube without the liquid paraffin dripping out.

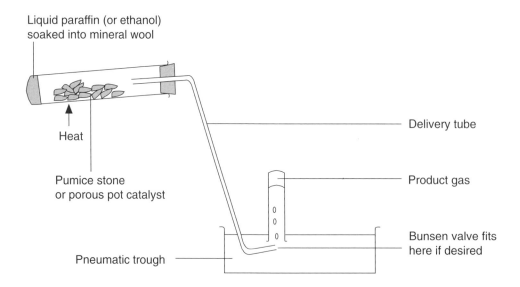

Fig. 2 Apparatus for cracking paraffin

THE ROYAL
SOCIETY OF
CHEMISTRY

Clamp the boiling tube so that the mouth is tilted very slightly upwards and pack the middle of the tube with pieces of pumice stone or broken porous pot. Clamp the tube so that the clamp is as close to the mouth as possible to avoid possible overheating of the clamp. Fit the delivery tube so that it dips into the water in the trough and fit a Bunsen valve if desired.

Hold a test-tube over the end of the delivery tube to collect gas. An assistant will be useful to manipulate the gas collection tubes, change them and cork them as necessary.

Heat the pumice stone (or porous pot) strongly with the hottest part of a roaring Bunsen flame for several seconds. Then flick the flame onto the mineral wool for a few seconds to vaporise some of the paraffin then return the flame to the pumice stone. Continue with the Bunsen burner heating the mineral wool for about one second out of every ten and the pumice stone for the other nine. Collect six test-tubes of gas. Discard the first two which will contain some air and cork and retain the other four.

Throughout, it is essential that the burner is not removed from the boiling tube or water will suck back up the delivery tube and possibly crack the boiling tube. If suckback begins, continue to heat strongly or remove the delivery tube from the water.

Some oil will be seen floating on the surface of the water in the trough. This will contain unchanged liquid paraffin that has distilled over and hydrocarbons with chains greater than C_5.

When six tubes of gas have been collected, remove the delivery tube from the water and stop heating.

Test the tubes of gas as follows.

1. Pass them round the class so that the students can cautiously smell the gas.

2. Uncork the tube and hold a lighted taper in its mouth. The gas will burn.

3. Add about 1 cm depth of bromine water to the test-tube. Re-cork and shake. The bromine water will be decolorised.

4. Add about 1 cm depth of acidified potassium permanganate solution to the test-tube. Re-cork and shake. The solution will be decolorised, possibly leaving a brown coloration of manganese dioxide.

The latter two tests indicate that the product has a carbon-carbon double bond.

For dehydrating ethanol

Follow exactly the same method as above but use ethanol instead of the liquid paraffin. In this case no oil will be seen on the surface of the water in the trough.

Teaching tips

The students will need to be familiar with the permanganate and bromine water tests for a carbon-carbon double bond.

Ball and stick molecular models will be useful for modelling both the cracking and the tests.

For the cracking of the paraffin point out that the product, being a gas, must have shorter chain lengths than the starting material.

For the dehydration of ethanol point out that the only likely product is ethene.

THE ROYAL
SOCIETY OF
CHEMISTRY

Theory

For the cracking of the paraffin
Simple cracking of a hydrocarbon produces two shorter chains, one of which is an alkene, for example, cracking decane could give heptane and propene.

For the dehydration of ethanol
The ethanol is dehydrated:

$$CH_3CH_2OH(g) \rightarrow CH_2= CH_2(g) + H_2O(g)$$

In both cases although bromine in non-aqueous solutions adds across the double bond of, say ethene, to give 1,2-dibromoethane (a suspected carcinogen), in aqueous solution, the main product is 2-bromoethanol.

Acidified permanganate ions cleave double bonds to form two carbonyl compounds while alkaline permanganate produces a diol.

Safety

Wear eye protection.

Use a safety screen because there is the possibility of the boiling tube shattering if suckback occurs.

It is the responsibility of teachers doing this demonstration to carry out an appropriate risk assessment.

THE ROYAL
SOCIETY OF
CHEMISTRY

99. The cornflour 'bomb'

Topic

Combustion, rates of reaction. Also an interesting open day demonstration.

Timing

Less than five min.

Level

Introductory chemistry.

Description

Cornflour is sprayed into the flame of a candle that is burning inside a catering-size coffee tin with the lid on. The resulting rapid combustion produces a small explosion that blows the lid off the tin.

Apparatus

▼ One 500 g (catering size) coffee tin with lid.

▼ One small (2–3 cm diameter) funnel.

▼ One short length of glass tubing.

▼ Small rubber bung with one hole.

▼ Short piece of candle.

▼ Rubber bulb-type pipette filler (about 50–100 cm³).

▼ Eight 1 cm x 1 cm x 1 cm Tillich bricks (wooden or plastic cubes) (optional).

▼ Safety screen.

Chemicals

The quantities given are for one demonstration.

▼ A few grams of dry cornflour.

Method

Before the demonstration

Make up the apparatus shown in the figure. Make a suitably sized hole near the base of the coffee tin to take the one-holed rubber bung. Insert the stem of the funnel into the hole in the bung so that the funnel is angled slightly upwards. Connect a short length of glass tube to the outside of the hole in the bung. Fit a bulb-type pipette filler to the other end of the glass tube. Place a short length of candle inside the coffee tin and stick it down with a little molten wax.

This apparatus can be improvised in a number of ways but it is important that the resulting apparatus can blow a cloud of cornflour into the candle flame when the pipette filler is squeezed. Some prior experimentation and adjustment may be needed.

The demonstration

Fill the funnel with cornflour, light the candle and quickly fit the tin lid. Quickly (before the candle goes out) give the pipette filler a rapid squeeze to blow cornflour

THE ROYAL
SOCIETY OF
CHEMISTRY

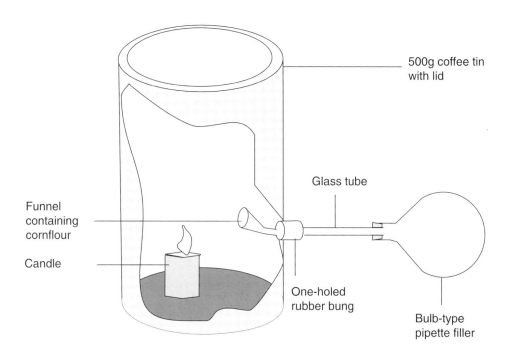

Cornflour bomb

into the candle flame, keeping your head well back. The resulting rapid combustion of the finely-divided cornflour blows the lid off the tin.

Teaching tips

Tell students about the explosion risks of powders such as sawdust from sanders and coal dust explosions in mines. The latter can be contrasted with the difficulty often encountered in igniting large lumps of coal when lighting a fire.

To illustrate that dividing solids up increases their total surface area, use eight 1 cm Tillich bricks. Make these into a 2 x 2 x 2 cube and show that the surface area is 2 x 2 x 6 = 24 cm². Now break the cube into its eight constituent bricks and show that the total surface area is now 1 x 1 x 6 x 8 = 48 cm². Also, shading the outside of the 2 x 2 x 2 cube with chalk before breaking it up shows that more surface is exposed when the cube is broken up because each of the smaller cubes will have three shaded and three unshaded faces.

This demonstration is a good illustration of energy changes in chemical reactions and that some compounds have a lot of energy 'locked up' in them. Ask students to identify the types of energy involved as the lid flies off. They should suggest heat, light, sound, kinetic and potential (the lid at the top of its trajectory). The idea of activation energy could be introduced to a suitable class.

Theory

Cornflour, which is a carbohydrate, burns rapidly because of its high total surface area to volume ratio which allows oxygen in the air to get at it easily.

Extensions

Try other finely divided combustible powders such as custard powder, flour or lycopodium powder.

THE ROYAL
SOCIETY OF
CHEMISTRY

Safety

Wear eye protection.

Use a safety screen between the apparatus and the audience.

Keep your head out of the way of the lid!

It is the responsibility of teachers doing this demonstration to carry out an appropriate risk assessment.

THE ROYAL
SOCIETY OF
CHEMISTRY

100. The oxidation of ammonia

Topic

Industrial chemistry, catalysis, reactions of ammonia.

Timing

About ten min depending on how many catalysts are attempted.

Level

Pre-16.

Description

Oxygen is bubbled into aqueous ammonia solution and a heated spiral of platinum wire is held in the gas mixture. The wire glows red hot and brown nitrogen dioxide and white fumes of ammonium salts can be seen in the flask.

Apparatus

▼   One 500 cm³ conical flask. Select one with as wide a mouth as possible.

▼   One length of glass tube, slightly longer than the height of the flask.

▼   One glass rod whose end has been bent into a hook.

▼   Bunsen burner.

▼   Access to fume cupboard.

Chemicals

The quantities given are for one demonstration.

▼   100 cm³ of **880 ammonia solution**.

▼   Access to an **oxygen** cylinder with appropriate regulator.

▼   Length of rubber tubing to connect the oxygen cylinder to the glass tube.

▼   About 25 cm of platinum wire (about 22–26 swg) but neither the length nor the thickness is critical.

▼   Similar lengths of wires of other materials in different thicknesses as available. For example copper, nichrome, iron, hecnum (constantan). **NB** hecnum and constantan are alternative names for an alloy of 55 % copper, 44 % nickel and 1 % manganese.

Method

The demonstration

Working in a fume cupboard, pour about 100 cm³ of 880 ammonia solution into the conical flask. Attach the glass tube to the oxygen cylinder and dip the glass tube into the ammonia. Make a spiral of platinum wire by winding a length of it around a pencil. Wind the spiral quite tightly so that each turn is close to the next. This slows down heat loss from the coil.

     Turn on the oxygen so that there is a rapid stream of bubbles through the ammonia solution. Hold the platinum spiral on the glass rod hook and heat it to red heat in the Bunsen flame. Lower the spiral into the flask of ammonia close to the

THE ROYAL
SOCIETY OF
CHEMISTRY

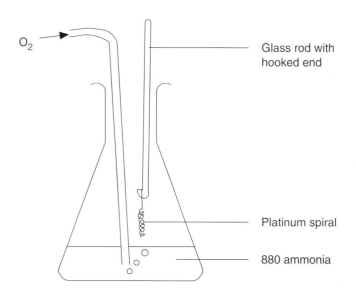

O_2

Glass rod with
hooked end

Platinum spiral

880 ammonia

The oxidation of ammonia

oxygen tube. The spiral which will have cooled to below red heat during the transfer will begin to glow red again as the ammonia is oxidised to nitrogen monoxide. This then reacts with further oxygen to give brown nitrogen dioxide. Brown fumes of this gas may be seen in the flask. Further reaction to ammonium nitrite and/or ammonium nitrate gives white fumes which may also be seen. If the flow of oxygen is sufficiently rapid, small, harmless explosions may occur which give yellow flames. These may become cyclic, *ie* the spiral glows, the mixture explodes, cooling the spiral which then heats up until a further explosion occurs.

Now repeat with the platinum replaced with a similar spiral of copper wire. This will glow but is less effective than platinum. Beware of overheating and melting the copper spiral.

Visual tips

A white background is essential if any brown fumes are to be detected.

Teaching tips

Point out that the reason that the spiral glows is that the oxidation of ammonia is an exothermic process. The faster the reaction proceeds, the faster heat is produced on the catalyst surface and if this is faster than the spiral can radiate it away, the spiral will heat up.

Theory

The reaction for the oxidation of ammonia under these conditions (*ie* solid platinum catalyst, moderate temperature and excess oxygen) is:

$$4NH_3(g) + 5O_2(g) \rightarrow 4NO(g) + 6H_2O(l) \; \Delta H = -909 \; kJ \; mol^{-1}$$

rather than

$$4NH_3(g) + 3O_2(g) \rightarrow 2N_2(g) + 6H_2O(l)$$

THE ROYAL
SOCIETY OF
CHEMISTRY

This is followed by

$$2NO(g) + O_2(g) \rightleftharpoons 2NO_2(g)$$

and

$$2NO_2(g) + H_2O(g) \rightarrow HNO_2(aq) + HNO_3(aq)$$

These acids react with ammonia to give ammonium salts.

The reactions are similar to those which occur in the Ostwald process for the manufacture of nitric acid from ammonia.

Extensions

Investigate the effect of changing variables such as

▼   thickness of wire;

▼   length of wire;

▼   material of wire;

▼   pitch (number of turns per unit length) of the spiral; and

▼   oxygen flow rate.

How systematically these can be done will depend on the availability of different wires. Platinum is very effective, copper and hecnum work satisfactorily while iron and nichrome do not appear to work at all. This could be done as an assisted demonstration, with students bringing various catalysts up to the fume cupboard to try them out. In each case the effectiveness of the catalyst can be judged by how brightly the spiral glows.

Investigate the effect of increasing or decreasing the oxygen flow.

Further details

The demonstration can be done with a platinum catalyst and air instead of oxygen. Lower the heated platinum spiral into the vapour above the ammonia and shake to admit more air. The spiral will begin to glow.

Chemically generated oxygen could be used if a cylinder is not available, but the flow rate is less easy to control.

Safety

Wear eye protection.

It is the responsibility of teachers doing this demonstration to carry out an appropriate risk assessment.